ENVIRONMENTAL TECHNOLOGY, ASSESSMENT AND POLICY

ELLIS HORWOOD SERIES IN ENVIRONMENTAL SCIENCE
Series Editor: R. S. SCORER, Emeritus Professor and Senior Research Fellow in Mathematics and Environmental Technology, Imperial College of Science and Technology, University of London

A series concerned with nature's mechanisms — how earth and the species which inhabit it fit together into a dynamic whole, and the means by which evolution has taught them to survive.

We are *not* primarily concerned to exploit the environment to human advantage, although that may happen as a result of understanding it.

We are interested in the basic nature of the physical world, the special forms that it takes on earth, the style of life species which exploit special aspects of nature as well as the details of the environment itself.

ATMOSPHERIC DIFFUSION, 3rd Edition
F. PASQUILL and F. B. SMITH, Meteorological Office, Bracknell, Berks
LEAD IN MAN AND THE ENVIRONMENT
J.M. RATCLIFFE, Visiting Scientist, National Institute for Occupational Safety and Health, Cincinnati, Ohio
THE PHYSICAL ENVIRONMENT
B. K. RIDLEY, Department of Physics, University of Essex
CLOUD INVESTIGATION BY SATELLITE
R. S. SCORER, Imperial College of Science and Technology, University of London
GRAVITY CURRENTS: In the Environment and the Laboratory
JOHN E. SIMPSON, Department of Applied Mathematics and Theoretical Physics, University of Cambridge

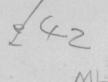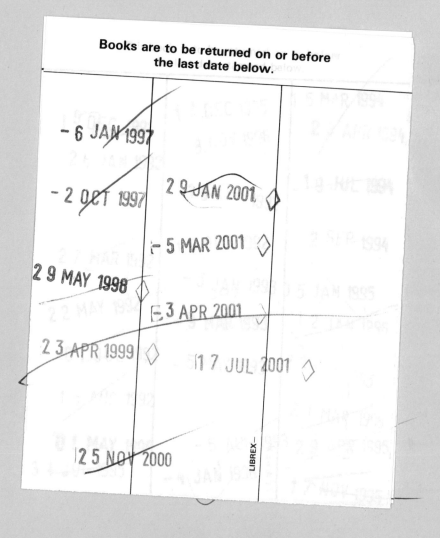

ENVIRONMENTAL TECHNOLOGY, ASSESSMENT AND POLICY

Editors:

R. JAIN, M.P.A., Ph.D., M.S., B.S., FASCE
Chief, Environmental Division, USA-CERL, Illinois

A. CLARK, B.Sc., Ph.D.
formerly of the Department of Civil Engineering
Imperial College of Science and Technology

ELLIS HORWOOD LIMITED
Publishers · Chichester

Halsted Press: a division of
JOHN WILEY & SONS
New York · Chichester · Brisbane · Toronto

First published in 1989 by
ELLIS HORWOOD LIMITED
Market Cross House, Cooper Street,
Chichester, West Sussex, PO19 1EB, England
The publisher's colophon is reproduced from James Gillison's drawing of the ancient Market Cross, Chichester.

Distributors:

Australia and New Zealand:
JACARANDA WILEY LIMITED
GPO Box 859, Brisbane, Queensland 4001, Australia

Canada:
JOHN WILEY & SONS CANADA LIMITED
22 Worcester Road, Rexdale, Ontario, Canada

Europe and Africa:
JOHN WILEY & SONS LIMITED
Baffins Lane, Chichester, West Sussex, England

North and South America and the rest of the world:
Halsted Press: a division of
JOHN WILEY & SONS
605 Third Avenue, New York, NY 10158, USA

South-East Asia
JOHN WILEY & SONS (SEA) PTE LIMITED
37 Jalan Pemimpin # 05–04
Block B, Union Industrial Building, Singapore 2057

Indian Subcontinent
WILEY EASTERN LIMITED
4835/24 Ansari Road
Daryaganj, New Delhi 110002, India

© 1989 R. Jain and A. Clark/Ellis Horwood Limited

British Library Cataloguing in Publication Data
Environmental technology, assessment and policy.
1. Environment. Pollution of control measures
I. Jain, R.   II. Clark, A.
628.5

Library of Congress card no. 89-15423

ISBN 0-7458-0687-2 (Ellis Horwood Limited)
ISBN 0-470-21494-5 (Halsted Press)

Typeset in Times by Ellis Horwood Limited
Printed in Great Britain by The Camelot Press, Southampton

COPYRIGHT NOTICE
All Rights Reserved. No part of this publication may be reproduced, stored in a retrieval system, or transmitted, in any form or by any means, electronic, mechanical, photocopying, recording or otherwise, without the permission of Ellis Horwood Limited, Market Cross House, Cooper Street, Chichester, West Sussex, England.

Table of contents

Preface 7

Part 1 Introduction
  1.1 Introduction 11

Part 2 Air pollution control
  2.1 Regulatory and technical measures for air pollution control in the
       Federal Republic of Germany 17
       H.-J. Oels
  2.2 Abatement technologies for air pollutants 42
       J. J. Stukel
  2.3 Economic issues in the control of air pollution 61
       I. M. Torrens

Part 3 Water treatment
  3.1 Megatrends in water treatment technologies 77
       F. Fiessinger, J. Mullevialle, A. Leprince and M. Wiesner
  3.2 Emerging technologies in the water supply industry .. 84
       J. F. Manwaring
  3.3 Trends in water treatment technology: disinfection, oxidation,
       and adsorption 97
       V. L. Snoeyink
  3.4 Water treatment research and development: a European viewpoint .. 114
       T. H. Y. Tebbutt

Part 4 Wastewater treatment
  4.1 Research needs for wastewater treatment and management of
       resulting residues 133
       R. I. Dick
  4.2 Wastewater treatment with coagulating chemicals 144
       H. H. Hahn

	4.3	Wastewater treatment by fixed films in biological aerated filters . . . . 170 J. Sibony

Part 5 Hazardous/toxic waste

	5.1	Technologies for treatment and management of selected hazardous/toxic wastes . . . . 179 L. W. Canter
	5.2	Hazardous waste: status and future trends of treating technologies. . . 198 F. J. Colon and J. H. O. Hazewinkel

Part 6 Environmental policy and regulations

	6.1	The EEC directive on environmental assessment and its effect on UK environmental policy . . . . 225 B. D. Clark
	6.2	UK pollution control — legal perspectives . . . . 237 Richard Macrory
	6.3	Evolution of technology strategies in environmental policy formulation . . . . 251 David H. Marks and Deborah L. Thurston
	6.4	Environmental regulation formulation and administration . . . . 257 R. K. Jain and D. Robinson

Appendix — List of participants 270

Index 274

Preface

The University of Cambridge, in cooperation with the US Army Research Development and Standardization Group — UK, conducted a workshop on Environmental Technology Assessment and Policy in April 1985 in Cambridge. A selected group of distinguished scientists and engineers from Western Europe and the United States of America were invited to this workshop. Attenders presented position papers and developed a review and critique of these papers for the following five major topic areas:

- Air pollution control
- Water treatment
- Wastewater treatment
- Hazardous/toxic waste
- Environmental policy and regulations

This book includes selected papers for each of the five areas mentioned above.

Workshop directors were Dr P. W. R. Beaumont (University of Cambridge, England), Dr R. K. Jain (US Army Corps of Engineers, USA), and Dr R. S. Engelbrecht (University of Illinois, USA).

Dr J. J. Stukel provided considerable assistance during the planning stages of the workshop. Dr A. Clark and Mr A. Baillie expended considerable effort in organizing and editing the workshop material to produce this book.

The book was made possible by the efforts of many participants in the conference who generously provided their time and support. Workshop participants and sponsor contributions and support are gratefully acknowledged.

Part 1
Introduction

1.1

Introduction

In industrialized countries, incorporation of environmental considerations in the planning and decision-making processes of government and industry has become a standard practice. Considerable new legislation has been promulgated in industrialized countries to further strengthen the environmental requirements. Protecting the environment, of course, has required countries to incur considerable costs.

Environmental regulations impose three kinds of cost on industry. First, there are the administrative costs of government agencies; second, there are direct compliance costs by industry in order for them to meet the environmental regulatory requirements; and, third, indirect costs are incurred because environmental regulations are likely to induce industry to seek less than optimal location for their plants and often encourage industry to maintain older, less economical plants and equipment.

The largest cost to industry deals with direct compliance cost. Statistics from the US Department of Commerce indicate that approximately 270 billion (thousand million) dollars in constant 1983 dollars were spent on water pollution control from 1972 through 1982 [1]. The US President's Council on Environmental Quality (CEQ) estimates that the US spent 36.9 billion dollars in 1979 to comply with federal environmental protection regulations alone. This expenditure of 36.9 billion dollars for the USA amounted to approximately 1.5% of GNP in 1979 [2]. Other Western European countries are spending resources at a similar level of environmental protection.

In addition to the costs incurred for environmental protection, concern about the way the regulations are enforced has been an important topic of discussion. It is generally felt that environmental pollution control which involves various economic incentive schemes might be more cost effective than the command-and-control type approach that is normally practised in most industrialized countries.

Economic incentives that focus on developing a system of effluent fees is one approach. Under this programme, each polluting industry will be required to pay a fee or tax for each unit of pollutant discharged. As a result, both the producer (through the tax) and the consumer (through subsequent price increases) are forced to internalize the pollution externality. This has the advantage of forcing the manufacturer to either change his level of production or adopt some pollution abatement technology. It also forces the consumer to decide whether to consume less of the goods or service. Clearly, technical problems in implementing such a strategy exist. Other suggestions focus on negotiable effluent discharge permits. These are not much different from the effluent charges, except that the fees for such permits will be determined by some type of market mechanism. Under these schemes, area-wide concentration limits can be set for various types of pollutants. Then permits that specify certain pollution limits are assigned to the existing firms.

Other innovative suggestions such as regulatory budget have been proposed (this is further described in the chapter by Jain and Robinson). All these economic incentive schemes are designed to reduce many inefficiencies that exist in the present pollution control approach. It is generally believed that such incentives can provide stimulation for further development of technology and, in fact, substantially reduce the pollution control costs to industry and ultimately to the consumer.

There are, of course, considerable benefits that society derives from environmental protection. Since benefits from environmental protection and enhancement are normally not marketable, quantifying these benefits is difficult and empirical evaluations can be widely variable. Despite these impediments, Ashford and Hill [3] were able to review and summarize many studies and to provide a range of benefits that can be attributed to environmental controls. As an example, they estimated that the US population, in terms of health costs and reduced loss of work, benefits from air pollution regulations by as much as $58.1 billion per year and water pollution regulations by as much as $10.4 billion per year. Additional benefits, such as less damage to physical structures, property values, wildlife and the natural environment, crops, and the natural resources cannot be quantified easily, but are considerable.

Benefits and costs of environmental protection and approaches to promulgation of environment regulations and policies for their implementation would naturally vary from one country to another. Also, technologies needed to address pollution control requirements in one country may not be completely applicable to another. Considering the emerging nature of environmental technology and the significant investment in environmental research and development industrialized nations are making, many benefits can be derived from the exchange of information on emerging technologies and related policies for their implementation. Consequently, a workshop was conducted at the University of Cambridge, England, in April, 1985, with the following objective: *to assess emerging technologies and research efforts dealing with environmental policies in Western Europe and the United States of America.*

The workshop covered five major areas: air pollution control; water treatment, wastewater treatment, hazardous/toxic waste; and environmental policies and regulations.

A number of position papers were prepared for each of the major topic areas. After the presentations of these papers, working groups discussed the material presented in the papers and provided summary in comments to critique and to expand upon the material presented in the position papers. These papers are organized in this book as follows.

Part 2 covers air pollution control and addresses the control of particulate emissions followed by control of gaseous emissions. For all these pollutants, the discussion briefly covers existing technologies and some emerging control technologies from the USA perspective. This chapter also includes a discussion of air pollution control regulations and technologies for controlling these pollutants with a specific reference to the Federal Republic of Germany (FRG). A summary of FRG air pollution control requirements is presented and specific technologies to control some of the pollutants are also included. In the case of air pollution control, as in other environmental protection, economic issues become quite important. With the primary focus on Organization for Economic Co-operation and Development (OECD) countries, emission standards for major air pollutants in these countries are presented. A brief discussion of technologies for control of regulated pollutants follows. Then, costs of controlling these pollutants are presented. A number of questions and issues associated with air pollution control are addressed:

- What is an appropriate level of control for different types of air pollution sources?
- How to treat new emission sources and existing sources?
- Who pays for increased controls?
- How to achieve the most economically effective results?
- What are the benefits of air pollution control?
- Finally, criteria for policy action are presented.

Part 3 covers water treatment and addresses emerging technologies in the water supply industry and innovative developments in all aspects of water supply field as reflected in Western European countries and the United States. In discussing megatrends in water treatment technologies, the effects of changes in (a) water quality standards, (b) technologies (such as electronics, biotechnologies, polymer sciences), and (c) opening of new markets for water treatment are presented. The presence of different types of organic compounds in the United States drinking water supplies has been a major force for change during the past decade. In addition, the widespread occurrence of volatile organic compounds (VOC) in groundwater has led to more frequent application of organic compound removal technologies. Focusing on organic compounds, (a) disinfection and chemical oxidation practice; and (b) adsorption are discussed.

Finally, water treatment research and development with the European perspective are included.

Part 4 covers wastewater treatment and includes discussion of traditional contaminants (for example, suspended solids removal, removal of biodegradable organic compounds, destruction of pathogenic organisms, and nutrient removal), microcontaminants (for example, heavy metals and organic microcontaminants) and sludge treatment, utilization and disposal. In addition, this chapter provides an extensive discussion of wastewater treatment using coagulating chemicals and wastewater treatment using fixed films in biological aerated filters.

Part 5 covers hazardous/toxic waste. Hazardous/toxic waste treatment and maanagement is a complex and emerging area of concern for the industrialized nations. Covered in this chapter are waste reduction technologies, waste treatment and disposal technologies, and extensive discussion on specific technologies. An extensive discussion of status and future trends of hazardous waste technologies is also presented.

Part 6 covers environmental policy and regulations. Included in this part are: (a) the European Economic Community directive on environmental assessment and its effect on UK environment policy; (b) legal perspectives on the UK pollution control; (c) evolution of technology strategies in environmental policy formulation; and (d) environmental regulation formulation and their administration.

The list of participants, along with their affiliations, is provided in the Appendix. The book is indexed to provide expedient access to topics discussed.

REFERENCES

[1] Conservation Foundation, *State of the Environment: An Assessment at Mid-Decade*, 1984.
[2] *Environmental Quality*, Eleventh Annual Report of the Council on Environmental Quality, U.S. Government Printing Office, Washington, D.C., 1980.
[3] Ashford, N. and Hill, C. 'The Benefits of Environment, Health, and Safety Regulation'. Prepared for the Committee on Government Affairs, U.S. Senate, 96th Congress, 2nd Session (25 March 1980).

Part 2
Air pollution control

2.1

Regulatory and technical measures for air pollution control in the Federal Republic of Germany

H.-J. Oels
Federal Environmental Agency, Bismarkplatz 1, D-1000 Berlin 33, FRG

1. INTRODUCTION

New types of damage to forests have occurred since the mid-1970s in the Federal Republic of Germany (FRG) and have been spreading at an alarming rate. According to the last survey, in 1984, 50% of German forests were affected by more or less severe damage. If the present course of disease is sustained the destruction of whole areas must be feared for medium altitude mountains.

Despite the present lack of unequivocal scientific evidence, the indications are that air pollutants — either alone, or in combination with other factors — play a crucial part in causing and determining the extent of the damage. There are also alarming reports about the acidification of aquatic systems caused by air pollution. Moreover, air pollution is held responsible for damage to buildings and cultural monuments.

There is therefore a need for more rigorous air pollution control measures not only to protect human health but also to safeguard our natural resources. Obviously, air pollution control in Europe is an international problem because of long-range transboundary transportation of air pollutants. On the other hand, independent national efforts in the field of air pollution control can encourage neighbouring countries to follow suit, and this can promote international cooperation. The FRG hopes that her national and international activities to improve the environmental situation are understood in this sense by other countries.

2. REGULATIVE MEASURES

2.1 Federal emission control law

The central legal instrument for air quality control policy in the FRG is the Federal Immission Control Law (BImSchG) of March, 1974 [1]. This law is aimed at:

— safeguarding an environment that man needs as a basis for a healthy and dignified life;
— protecting animals, plants and material goods, as well as soil, air and water from the detrimental influences of human activities.

These aims mean that beyond averting immediate threats and removing damage that has already been caused, considerable efforts have to be made in order to prevent environmental damage from occurring in the future (the principle of prevention).

Two main strategies are incorporated in the BImSchG.
(1) Limitation of immissions: ambient air quality standards are established so that tolerable emission loads are not exceeded by the development of new emitters or the extension of existing emitters.
(2) Limitation of emissions: The defining of emission standards determined by the best available technology serves to implement all technically possible and appropriate means to reduce emissions in every individual case, independent of the specific local ambient air quality. The costs involved must be borne by the polluter.

These strategies are implemented by installation-, area- and product-related measures. All facilities with relevant emissions — listed in a special ordinance (4. BImSchV) [2] — are subject to a licensing procedure. Limitation of emissions, regulation of stack heights, and definition of monitoring and control systems are considered the licensing procedure. The BImSchG authorizes state governments to establish air quality plans for highly polluted areas, with the aim of a long-term improvement of the situation, and to define clean air areas for conservation of the positive situation. A typical product-related measure is represented by the Third Ordinance implemented by the BImSchG (3. BImSchV) [3]. By this ordinance the sulphur content of gas oil (light fuel oil and diesel fuel) has been limited to a maximum of 0.3% in the FRG since January 1979. A further reduction of the sulphur content to 0.15% is now under discussion.

A specific problem in air quality control is older plants which are no longer in line with the advanced emission reduction technology but which are still allowed to operate because of their former licence. The BImSchG provides some possibility of imposing further requirements in accordance with the advanced state of reduction technology. But these additional impositions can only be accomplished if they are economically feasible. This restriction has often prevented retrofitting. To facilitate this difficult procedure the Federal Minister of the Interior has now submitted a bill to amend the BImSchG in which the requirement of economic feasibility in particular is weakened. Another important objective is to realize a dynamic state of reduction technology which is adjusted to technical advancements from time to time. The most important ordinances and regulations for the implementation of the BImSchG are the First Ordinance (1. BImSchV) [4], the Thirteenth Ordinance (13. BImSchV) [5] and the Technical Instruction for Air Pollution Control [6].

2.2 First ordinance implementing BImSchG (1. BImSchV)

The 1. BImSchV covers all smaller firing installations that are not subject to a licensing procedure. There are about 5 million oil-fired and an estimated 0.5 million solid-fuelled facilities with heat inputs <1 MW, as well as about 2.9 million gas-fired

Ch. 2.1] **Regulatory and technical measures for air pollution control** 19

facilities with a heat input up to 100 MW. All facilities, except the larger gas firings, are for room heating or warm water preparation.

The requirements which must be met refer to the stack losses (i.e. energy content in the flue gas), to certain fuel specifications, and to emission standards. Details of the requirements are given in Tables 1 and 2. The chimney sweep controls of the

Table 1 — Requirements of the 1.BImSchV. Stack losses for oil and gas fired installations

Thermal output[a] (kW)	Stack loss (%)[b]		
	Date of installation		
	Before 1.1.79	Since 1.1.79	Since 1.1.83
4–11	18	16	14
11–25	18	16	14
25–50	17	15	13
50–120	16	14	12
>120	15	13	11

[a] Measurement only at start of operation.

[b] Stack loss $=f\frac{t_A-t_L}{CO_2}$ where: t_A=flue gas temperature (°C); t_L=ambient air temperature (°C); CO_2=CO_2 concentration in flue gas (%); f=0.59 for light fuel-oil, 0.50 for liquid gas, 0.38 for coal-derived gas and forced draft burners, 0.35 for coal-derived gas and natural draft burners, 0.46 for natural gas and forced draft burners, 0.42 for natural gas and natural draft burners.

requirements are performed by yearly repeated measurements at all oil- and gas-fired facilities; hand fed coal- and wood-fired facilities are only controlled at their first start-up. The success of the ordinance is difficult to quantify because of the continuous fuel switch from coal to oil, and now to gas, during the last two decades in the heating sector. But the observed decreasing number of objections is an evident indication of the improvements achieved. A further tightening up of the regulations is being prepared.

2.3 Thirteenth ordinance implementing the BImSchG (13. BImSchV)

The 13. BImSchV (Ordinance on Large Firing Installations) came into force on 1 July 1983. The ordinance applies to all stationary combustion installations, predominantly boilers but also refinery furnaces, with a thermal heat input of at least 50 MW; for gas-only-fired facilities the ordinance applies to installations >100 MW. It should be mentioned that several firing units at one location form one facility if they are in close operational proximity. Thus even smaller units are covered by the ordinance if the sum of their heat inputs reaches 50 or 100 MW. For these installations, the ordinance gives emission limits for total dust and some heavy metals, sulphur oxides, nitrogen oxides, carbon monoxide and gaseous inorganic chlorine or fluorine compounds. Table 3 lists the most important emission standards for SO_2, NO_x and

Table 2 — Requirements of the 1. BImSchV. emission standards

Type of facility	Date of installation	Smoke No[c] (Bacharach)	CO_2 (% least)	Dust, soot, tar (mg/m^3 correlated to 12% CO_2)
Atomizing burner for liquid fuels[a]	before 10.1.74	3	7	—
	after 10.1.74	3	10	—
Vaporizing burner for liquid fuels >11 kW[a,b]	before 1.1.79	3	—	—
	after 1.1.79	3	8	—
Solid fuelled >22 kW hand filled	—	—	—	150 (wood fired: 300)
Solid fuelled >22 kW Mech. filled	—	—	—	300
Solid fuelled <22 kW				Low smoke operation must be guaranteed by (a) use of low smoke fuels or (b) installation of a *Universaldauerbrenner* (Special stove)

[a] Only light fuel oil according to DIN 51 603 part 1 is permitted
[b] Facilities <11 kW or not controlled have to meet certain DIN construction standards
[c] Additionally, an acetone test must show no oil derivatives.

Table 3 — Emission standards for large firing installations according to the 13. BImSchV and the UMK decision (in mg/m^3)

Capacity (MW)	Kind of installation	Solid fuels (Ref.O_2=6% wet bottom, 5% grate/FBC:7%)			Liquid fuels (Ref. O_2=3%)			Gas fuels[c] (Ref. O_2=3%)
		NO_x	Dust	SO_2	NO_x	Dust	SO_2	NO_x
50–100	New	400	50	2000	300	50	1700	—
	Old	650 (1300 wet bottom)	125 (80 lignite)	2000	450	50–100	1700	—
100–300	New	400	50	2000[a]	300	50	1700[a]	200
	Old	650 (1300 wet bottom)	125 (80 lignite)	2,000[a]	450	50	1,700[a]	350
>300	New	200	50	400[b]	150	50	400[b]	100
	Old	200	125 (80 lignite)	400[b]	150	50	400[b]	100

[a] Maximum permissible sulphur emission rate 40% or light fuel oil with 0.3% sulphur.
[b] Maximum permissible sulphur emission rate 15% or light fuel oil with 0.3% sulphur.
[c] SO_2 and particulate emission standards correlate to usual sulphur and dust contents of fuel gases.

particulate matter. The standards are closely related to detailed measuring and monitoring provisions.

In most cases control devices with continuous monitoring are required. The emission concentrations are averaged during a time period of 30 minutes. The emission standard is defined by three requirements.

— the daily 24 h average of all half-hour values must be below the standard
— 97% of all half-hour values during a year must be below 120% of the standard
— 100% of all half-hour values during a year must be below 200% of the standard.

As significant emission reductions over a relatively short time can only be achieved by drastic reduction measures at existing plants, the ordinance also includes emission limits for these plants. As a rule all existing plants that are covered by the ordinance must meet these standards by 1 July 1988, or they must end their operation by 1 April 1993 at the latest, depending on their yearly operating hours. For instance, retrofitting with flue gas desulphurization (FGD) by 1988 is required for oil and coal-fired facilities >300 MW if the remaining operating time exceeds 30 000 h. The use of low sulphur fuels (<1% S) is sufficient if the remaining operating time is shorter. For medium-sized facilities (100–300 MW) a longer time period (until 1993) is provided for the obligatory installation of 60% desulphurization. As described above, the relevant plant size is defined by the total installed capacity at one location.

In practical terms, this means that for 90% of the total capacity of all power plants in the FRG, FGD has to be installed. Old plants that will continue operation for less than 10 000 hours have to comply with their operating permits as licensed. A time span of one year until 1 July 1984, was granted to operators to decide about which installations should be closed down and which ones were to be retrofitted.

With respect to the nitrogen oxide emission standards, new technical developments had to be taken into account since the drafting of the ordinance. The Ordinance on Large Firing Installations prescribes NO_x emission standards that can be achieved relatively cost-effectively by improving combustion processes (low NO_x-burners and boiler modifications). Furthermore, a clause on dynamic adaption requires the application of any new measure to reduce further emissions.

This gave rise to some uncertainties and therefore on 4 April 1984, the Conference of the Federal and State Ministers for the Environment (UMK) agreed to replace this clause by stringent, clearly quantified standards. According to their decision, flue gas denitrification must be installed in new and existing installations >300 MW that will operate longer than 30 000 hours. The new NO_x-standard has to be met as soon as possible. All indications are that 1990 is an appropriate date. Further details are given in Table 3.

As a consequence of the 13. BImSchV a drastic reduction of SO_2 and NO_x emissions from stationary sources is expected. Table 4 gives a comparison between the 1982 and the estimated 1995 emission situation. The calculations show a reduction of the annual SO_2 emissions of about 1.5 million tons and of the annual NO_x emissions of about 0.6 million tons. This is a 50% SO_2 reduction and a 40% NO_x reduction for stationary sources. It should be mentioned that most of the reductions will be already achieved by 1990.

Based on a survey carried out by the Federal Environmental Agency among

Table 4 — Emission situation in the FRG 1982 and — as a result of the ordinance of large firing installations — in 1995 (in 1000 tons/y)

Sector	SO$_2$ 1982	SO$_2$ 1995	NO$_x$(NO$_2$) 1982	NO$_x$(NO$_2$) 1995	Particulates 1982	Particulates 1995
Power and district heating plants	1860	410	860	260	150	70
Industry	760	670	430	370	420	400
Household and other small consumers	280	no impact	120	no impact	60	no impact
Transportation	100	no impact	1690	no impact	70	no impact
Total	3000	(1460)^a	3100	(2400)^a	700	(600)^a

^{a} Quantities are being further changed by other regulations, in particular in the transportation sector.

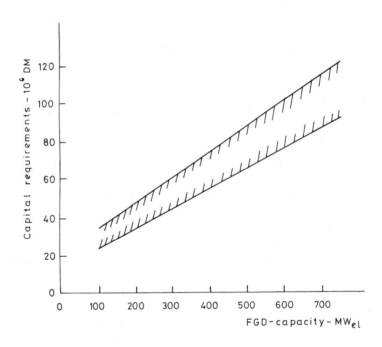

Fig. 1 — Capital requirements for the purchase of the desulphurization facility as a function of FGD capacity (basis 1982)

producers and suppliers of flue gas desulphurization plants in the FRG, average cost functions for FGD were determined. Figure 1 gives the capital cost requirements for the purchase of FGD units as a function of desulphurized capacity. Additional capital costs are needed to cover site-specific measures (foundation, housing, etc.) and interest payments during construction [7]. These additional expenditures amount to approximately 20% and, in case of difficult retrofit situations, may be as high as 50%.

Table 5 shows the desulphurization investment programme resulting from the

Table 5 — Desulphurization investment programme resulting from the implementation of the Ordinance on Large Firing Installations

Energy source	1982 Capacity	1995				
		Capacity[a]	Work[a]	Retrofit FGD	New FGD	Cumulative Investment
	(GW_{el})	(GW_{el})	(TWh)	(GW_{el})	(GW_{el})	(10^6 DM)
Hard coal (including existing FGD capacity)	30.4 (2.4)	35.5	146.8	20	13	8600
Lignite	13.7	13.7	42.2	11	2.7	3800
Fuel oil (including existing FGD capacity)	13.8 (0.2)	7.8	9.2	1	—	300
Gas	14.6	10.2	32.8			
Nuclear energy	10.4	23.9	143.2	no SO_2 emissions		
Hydropower	6.5	6.7	19.4	no SO_2 emissions		
Total	89.4	97.8	443.6			12 700

[a] Estimation (middle version) by Prognos/Switzerland, 1984.

implementation of the ordinance. Based on these data the implementation of the Ordinance on Large Firing Installations requires a total investment of DM 13 000 million for the desulphurization of public and private power plants. The investment to be expected as a consequence of the desulphurization of refineries and other heat producers has not yet been estimated.

With respect to flue gas denitrification no detailed data on the type and extent of the necessary technical measures are yet available to undertake a macroeconomic estimate of the investment. As guiding figures, approximately 30–50% of the desulphurization costs can be assumed as investment costs for $DeNO_x$ plants in the FRG.

Adding all costs for SO_2 and NO_x reduction, the expenditures for improved dust precipitators and the installation of better emission monitors, the total investment required until 1995 amounts to about DM 20 000 million.

Total annual costs for capital services and operation of FGD units can be estimated to about DM 3100 million. Taking into account that the total electricity production in the FRG is estimated to be 450 TWh in 1995, and that more than half of this will be produced in fossil fuel fired power plants equipped with FGD, the desulphurization costs amount to about 0.013 DM/kWh. This corresponds to a total average rise in electricity prices of DM 0.007 per kWh, if electricity generation from all energy sources is considered. In relation to the present electricity price level of 0.25 DM/kWh for households and 0.13 DM/kWh for industrial users the price increase will be 2.5 and 5% respectively.

Including the costs of $DeNO_x$ plants and of other measures in consequence of the ordinance, an average increase in electricity production costs will amount to 0.02 DM/kWh, which corresponds to a rise in overall electricity prices of 0.01 DM/kWh.

2.4 Technical instruction for air pollution control (TA Luft)

The TA Luft is a general administrative regulation based on the BImSchG and contains immission and emission limits of particular importance for the licensing of industrial plants and the subsequent imposition of standards on existing plants. In particular, the TA Luft contains formulae for pollutant distribution calculations and for the determination of stack heights. A model for ambient air quality monitoring is introduced as well as a method for an ambient air quality prognosis, obligatory for the determination of the impact of every new plant being subject to a licensing procedure.

The emission provisions of the TA Luft as of 1974 were amended with effect from 1 March 1983. For the first time emission values for lead, cadmium and thallium were introduced, and emission standards for carcinogenic substances were tightened drastically (Table 6). Preparatory work is under way on a further amendment of the TA Luft to restrict emissions of airborne pollutants in accordance with the advanced state-of-the-art technology. This amendment will cover almost all areas of industry, in particular blast furnaces, steelworks, lead smelters, coking plants, cement works, chemical plants, firing installations with a capacity of less than 50 MW_{th}, which are not covered by the Ordinance on Large Firing Installations, as well as large-scale animal farms and carcass disposal plants.

Similar to the 13. BImSchV, the amended TA Luft is intended to contain a general regulation for existing plants: all existing plants have to be retrofitted to meet the emission standards of new plants within five years or they have to be closed down within the next ten years at the latest. The new version of this part of the TA Luft is to be finalized and brought into force before the end of 1985.

2.5 Other regulations

Fossil fuel combustion for energy production is the main origin of air pollution and here energy-saving measures are also pollution control measures. The largest part of the final energy consumption is for domestic use (heating and hot water). Therefore most of the valid energy saving regulations deal with this sector. Special attention should be paid to the Ordinance on Thermal Insulation of 1977, amended in 1982 [8]. The revised text introduces step-by-step requirements for the thermal insulation of new buildings and, to a certain extent, of existing buildings.

The Ordinance on Heating Installations of 1978, amended in 1982, should also be

Table 6 — Ambient air quality standards established by the TA-Luft in 1983[a]

Pollutants	Units	Standard Health protection		Standard Protection against harmful effects	
		IW 1[b]	IW 2[c]	IW 1[b]	IW 2[c]
Dust concentration (regardless of composition)	mg/m^3	0.15	0.30		
Lead and inorganic lead compounds in dust	μg/m^3	2.0	—		
Cadmium and inorganic cadmium compounds in dust	μg/m^3	0.04	—		
Chlorine	mg/m^3	0.10	0.30		
Hydrogen chloride (Calc. as Cl$^-$)	mg/m^3	0.10	0.20		
Carbon monoxide	mg/m^3	10	30		
Sulphur dioxide	mg/m^3	0.14[d]	0.40		
Nitrogen dioxide	mg/m^3	0.08	0.30		
Dust fall (non-hazardous dust)	g/(m^{3}d)			0.35	0.65
Lead and inorganic lead compounds in dust fall	mg/(m^{2}d)			0.25	—
Cadmium and inorganic cadmium compounds in dust fall	μg/(m^{2}d)			5	—
Thallium and inorganic thallium compounds in dust fall	μg/(m^{2}d)			10	—
Hydrogen fluoride and inorganic gaseous fluorine compounds (calc. as F$^-$)	μg/m^3			1.0	3.0

[a] TA-Luft is the 'Technical Instructions for Air Pollution Control'.
[b] Annual average (arithmetic mean).
[c] Short-term exposure (98% value of cumulative frequency).
[d] If the ambient SO$_2$ loading in an area is below 0.05 or 0.06 mg/m^3 (annual average), no new construction that would cause this level to be exceeded will be allowed (i.e. preservation of clean areas).

mentioned, which regulates in particular energy-saving construction details of heating plants [9]. Additionally, accompanying subsidy programmes have been introduced in order to accelerate the implementation of energy-saving measures.

As can be seen in Table 3 the biggest contribution to NO$_x$ emissions comes from motor vehicles. Therefore, in 1983, the federal government decided to adapt, parallel to the introduction of unleaded petrol, the emission standards valid in the

United States and the corresponding test methods used there from 1 January 1986. Compared to the present situation in the FRG the United States standards imply a 90% NO_x reduction and considerable reductions of CO and hydrocarbon (HC) emissions.

The FRG would have preferred concerted European action to introduce the new standards, but she is also determined to proceed alone if other countries cannot join at this time. This problem shows the importance of international cooperation. Organizations like OECD, ECE and EEC are appropriate platforms for international environmental activities. The FRG supports the work of these organizations in this field. The FRG is a signatory of the ECE Convention on Long-Range Transboundary Air Pollution, and has obliged herself — like others — to reduce the annual SO_2 emissions of 1980 by 30% by 1993.

At present the Council of the EEC is discussing the proposal of the EEC Commission on a Guideline of Air Pollution Controls for Large Firing Installations [10]. The main objective of the guideline is to introduce unique SO_2, NO_x and dust emission limits for all large firing installations in the EEC, and thus to achieve a Europe-wide reduction of emissions. The FRG has been striving to implement stringent emission standards for new and existing facilities as soon as possible. Unfortunately major disagreements between the EEC members have prevented realization of the guideline until now.

3. TECHNICAL MEASURES

The main efforts of the FRG in air pollution control are focusing on the reduction of the pollutants SO_2, NO_x, and dust. These pollutants are mainly emitted by combustion of fossil fuels in stationary and mobile sources. Technical measures for emission reduction are: the use of low-pollution fuels; the modification of combustion processes; and flue gas cleaning.

Pollution control by energy saving is a further indirect reduction method, but it will not be discussed in this chapter.

3.1 Reduction of SO_2 emissions
3.1.1 Desulphurization of fuels

SO_2 emissions from fossil fuel combustion are a direct consequence of the sulphur content of the fuel. Normally almost the total sulphur content of the fuel is converted to SO_2 during combustion. Only at the coal burning stage is sulphur capture in the ash possible. Usual retention percentages are 5% for hard coal and about 40% for Rhinish Lignite. The sulphur content of German hard coals normally varies between <1% and 2%, the two main types averaging at 1% and 1.5% respectively. Nearly all of the yearly mined 83×10^6 T hard coals pass through a physical coal cleaning, i.e., they are crushed and separated from mineral material by stratification and flotation methods. This separation simultaneously removes the pyritic sulphur in the coal. The organic sulphur in coal (0.7% average) cannot be removed by physical means. This is only possible by coal conversion processes like coal gasification or coal liquefaction. Up to now only pilot hard coal conversion plants have been operated, as economic aspects have prevented commercial plants from realization. For physical coal cleaning the cost situation looks better: normally the cleaning is performed to such an extent that the additional costs are offset by the benefits of the higher-grade coal such as reduced costs for coal transportation, coal handling, pulverizing, ash handling and disposal and boiler maintenance. For sulphur emission control the costs for physical coal cleaning are therefore regarded as almost zero. Efforts have been directed at improving mechanical processing in order to desulphurize hard coal as much as possible mechanically at the lowest possible expenditure. But the achievable coal desulphurization is insufficient to comply with the emission limit value of 400 mg/m^3 prescribed for firing installations >300 MW. Cost calculations indicate that in this case flue gas desulphurization is more favourable than an extension of the coal cleaning.

Rhinish lignite (125×10^6 T/y) contains only organic sulphur, and desulphurization by physical methods is therefore not possible. However, the low sulphur content of this fuel (*ca.* 0.4%) in conjunction with the high sulphur capture in the ash make Rhinish lignite less serious in respect to SO_2 emissions than hard coal. There are more problems with some smaller high sulphur lignite open castings in Lower Saxonia and Hesse. Like the Rhinish lignite, these coals are used in power stations without any coal cleaning. This makes a highly effective post-combustion treatment necessary.

Liquid fuels are normally desulphurized in a refinery process. Direct desulphurization takes place during normal distillation, the sulphur being concentrated in the residue (Table 7). A further desulphurization of the distillate fuel is possible by

Table 7 — Sulphur content of crude oil fractions

Fraction (Boiling range °C)	Crude oils							
	Low sulphur		Medium sulphur		High sulphur			
	Brega (0.21 wt%)		Arabian Light (1.70 wt%)		Arabian Heavy (2.85 wt%)		Kuwait (2.50 wt%)	
	Volume	Sulphur	Volume	Sulphur	Volume	Sulphur	Volume	Sulphur
Naphtha (C5/154°C)	24.0	0.1	20.0	0.1	15.0	0.1	17.0	0.1
Distillate (154/371°C)	41.0	0.1	39.5	0.8	32.0	1.1	35.0	1.0
Residue (371°C+)	32.5	0.4	39.0	3.1	49.5	4.5	45.5	4.3

Source: CONCAWE, report No 5/81, 'Direct desulphurization of residual petroleum oil — investments and operating costs'.

treating these oil fractions with hydrogen in the presence of a catalyst. This well-established process is capable of 90% sulphur removal.

A direct desulphurization of atmospheric and vacuum residues is technically feasible but not common in German and European refineries. The residues are usually blended with low sulphur distillates derived from the crude oil to meet certain fuel oil standards (indirect desulphurization). Due to the increasing demand for naphtha and light fuel oils, heavy fractions of crude oil are now often converted to lighter products which can be desulphurized by catalytic hydrogenation. In the future the remaining heavy fractions will be fired in plants where desulphurization of flue gases is required. Increasing attention has to be paid to heavy metals (V, Ni, etc.) in fuel oils. Like sulphur, heavy metals are concentrated in the residue. Therefore firing installations for oil that contains more than 12 p.p.m. Ni must take additional precautions to reduce the emission of heavy metals.

The use of heavy fuel oil in the FRG is greatly on the decline. In 1982 a total of 14 million T were consumed; a figure of 6 million T has been forecast for 1995. Light fuel oils (160–390°C) with a sulphur content of 0.3% constitute the majority of oil consumption. In the near future the sulphur content will be reduced to 0.15%.

Fuel gases are usually desulphurized to suppress pipeline corrosions. Sulphur elimination techniques with efficiencies up to 99% are well established. The standard process-combination for desulphurization of natural sour gas is a washing unit and Claus kiln, often followed by a tail-gas treatment unit. Every year about 900 000 T of sulphur are produced by natural desulphurization plants in the FRG.

3.1.2 Combustion modification processes

SO_2 emission control by combustion modification processes is usually by increasing the sulphur capture in the ash. For this purpose basic sorbents like lime (CaO) limestone ($CaCO_3$), limehydrate ($Ca(OH)_2$) or dolomite are either added to the fuel prior to combustion or injected separately into the firebox. The sorbent reacts with SO_2 to form solid sulphates which can be removed by dust filters. The desulphurization efficiency depends on many parameters, such as type and particle size of the sorbent, reaction temperature, mixing conditions, etc. The most common sorbents are $CaCO_3$ and $Ca(OH)_2$, the latter being more expensive but resulting in higher desulphurization efficiencies. Sodium-based sorbents are uncommon in Germany.

For low Btu brown coal, like the Rhinish lignite, the so-called dry additive procedure (TAV) has been developed: the sorbent, $CaCO_3$, is added to the fuel before the mills [11]. The low combustion temperature (<1100°C) and the good mixing provide a 40–60% desulphurization. The consumption of $CaCO_3$ is about 4 tonnes/tonne SO_2 removed. The advantages of TAV are low investments and easy operation. TAV operates at one 300 MW_{el} power station and will be installed at 16 further units, with a total capacity of 2300 MW_{el} during the next 1–2 years. This action is to be considered as an emergency desulphurization measure until more efficient wet FGDs are on the line at these units in 1988.

Coal firings with higher combustion temperatures like pulverized hard coal firings and grate firings for hard coal are less successful with sorbent/coal mixtures prior to combustion because of thermal inactivation of the sorbent. Only fluidized bed combustion (FBC) represents a low temperature (800–900°C) combustion technology where a sorbent addition prior to combustion can be applied for a SO_2 reduction. At usual atmospheric stationary FBCs desulphurization efficiencies of 50–70% are achieved if limestone is added at a molar ratio of about Ca/S=2. In the FRG, twelve stationary FBCs are installed with heat ratings from 0.5 to 125 MW and a total capacity of about 470 MW [12]. A much better desulphurization is achieved by the atmospheric circulating FBC. The first commercial plant, an 84-MW steam boiler, has been operating since 1983 with 90% desulphurization at a Ca/S molar ratio of 1.5 [13]. Two further circulating FBCs with heat ratings of 109 MW and 226 MW will start operation by the end of 1985.

Ch. 2.1] Regulatory and technical measures for air pollution control

The first commercial pressurized stationary FBC in the FRG is now under construction; the start-up of the 42 MW facility is planned for 1985. It is expected that the pressurized FBC will achieve performances comparable to the circulating FBC. Special sorbent injection systems for normal pulverized coal and grate firings have also been developed. These so-called direct desulphurization procedures (DEP) make special provisions to prevent the sorbent from thermal inactivation (sintering).

Figure 2 shows the design of a limestone injection multistage burner (LIMB)

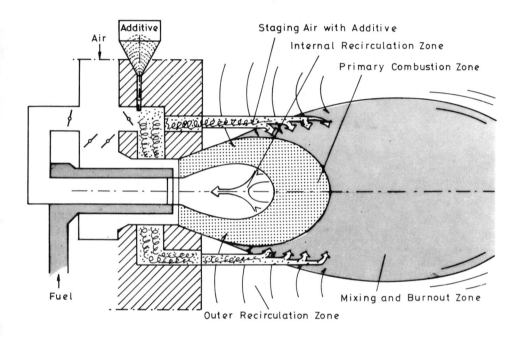

Fig. 2 — Direct desulphurization through additive injection around the SM burner.

which provides both SO_2 reduction by sulphur capture in the ash and low NO_x combustion by air staging [14]. The additives do not pass through the hot flame core, turbulences in the outer region of the flame ensure good mixing of the additive with the flue gas, and the whole residence time of the flue gas between burner and dust separator can be used as reaction time. Figure 3 shows some results from a 2.3-MW test combustion chamber. Limehydrate addition at molar ratios of about Ca/S=2 resulted in desulphurization rates of up to 50% [14]. The system is now installed in a 700 MW_{el} hard-coal fired power station for test trials and for further investigations; results are not yet published. Economic feasibility calculations show that DEP is especially interesting for smaller units and those with lower operating time because of the low investments necessary (*ca.* 10% of wet FGD).

A much simpler method is sorbent injection by overfire air nozzles. This procedure also combines low NO_x combustion with SO_2 reduction. First test trials

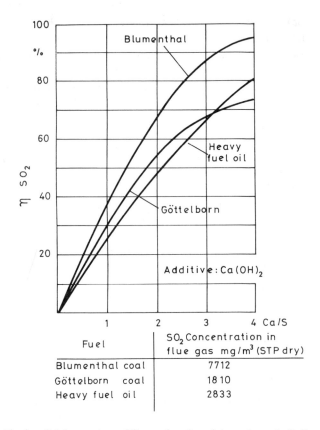

Fuel	SO₂ Concentration in flue gas mg/m³ (STP dry)
Blumenthal coal	7712
Göttelborn coal	1810
Heavy fuel oil	2833

Fig. 3 — Sulphur capture ηSO_2 as a function of the molar ratio Ca/S.

were carried out at the 300-MW_{el} hard-coal fired power station at Hamburg-Wedel last year. Start-up of commercial operation is planned for March/April 1985. The desulphurization rates achieved are comparable with the DEP process. Sorbent injection above the firebed is also feasible, for grate firings. After the successful conclusion of some tests several travelling grate facilities are now being equipped with sorbent injection devices for sulphur removal. Results are sensitive to the location and mode of injection. Adding $Ca(OH)_2$ at a molar ratio Ca/S=3 to the 1%-sulphur hard coal of a 64-T/h boiler resulted in a 26% desulphurization; injecting the same amount through the secondary air ports, a 40% desulphurization was achieved; injection by special lances 6 m above the grate gave the best result: a 60% SO_2 reduction.

The results reported with direct desulphurization are the basis for a new passus which is intended to be introduced into the revised TA Luft: in addition to a SO_2 emission concentration limit (coal: 2000 mg/m³–7% O_2; oil: 1700 mg/m³–3% O_2) for boilers <50 MW_{th}, there is notification to the licensing authority that direct desulphurization by sorbent injection is capable of a further 50% SO_2 reduction.

A specific problem of all sorbent addition systems is waste disposal: the quantity of the residue is multiplied and the composition is changed by sorbent treatment, limiting the utilization of the product for recycling. Moreover, underground water pollution by sulphates is possible if the residues are disposed of without great care.

Thus the further application of desulphurization by sorbent addition is closely linked to the problem of finding an ecologically and economically reasonable utilization of the residues.

3.1.3 Flue gas desulphurization

The ordinance on Large Firing Installations limits the SO_2 emission of coal- and oil-fired facilities >300 MW_{th} to 400 mg/m^3; at the same time the permissible sulphur emission rate is limited to ≤15%. The intention of this double limit is to save low sulphur fuels for smaller combustion units and to direct higher sulphur-containing fuels to facilities equipped with FGD.

This stringent SO_2 emission standard applies to about 95% of the installed coal- and oil-fired power station capacity in the FRG; it can only be met by flue gas desulphurization. Oil-fired power stations contribute only 3% to electricity production, and therefore most FGD activities focus on coal-fired units. By the end of 1984 a capacity of 4700 MW_{el} (4300 MW hard coal, 300 MW oil) was equipped with FGD, a further 5100 MW_{el} were under construction and 23360 MW_{el} in the planning phase. An estimate over all new and retrofitted public and industrial power stations results in a sum of about 50000 MW_{el} which will be desulphurized at the beginning of the 1990s.

The most common FGD plants in the FRG are based on wet lime/limestone scrubbing processes. Figure 4 shows a schematic drawing of a typical advanced lime/limestone FGD: downstream from the electrostatic precipitator the flue gas enters an absorber, where a slurry of limestone or limehydrate is sprayed into the gas to cool it down and to react with the SO_2 forming sulphite and sulphate. The flue gas then passes through a mist eliminator, and a heat exchanger raises its temperature before final emission to the atmosphere. The sulphite in the slurry is oxidized to sulphate (gypsum) by addition of oxidizing air. The gypsum is then separated from the liquid phase by special separators and centrifuges. The consumed limestone and water are continuously replaced. A particular feature of the flow scheme shown in Fig. 4 is the location of the fan downstream from the absorber in the cooled clean gas, instead of being located upstream of the heat exchanger in the hot dirty gas. This arrangement has the following advantages.

— Simple grouping of tubes and fan; less spacing problems with retrofits.
— Smaller fan size because of contracted gas volume (due to lower temperature).
— The compression caused by the fan raises the temperature of the gas and thus encourages evaporation of remaining droplets.
— Reversed gas slip in the heat exchanger from clean to dirty gas; thus a degradation of the desulphurization efficiency is excluded.

The main objectives of German FGDs are:
— to achieve a high overall desulphurization efficiency and a high availability at minimal costs;
— to evade disposal problems by producing a marketable by-product; and

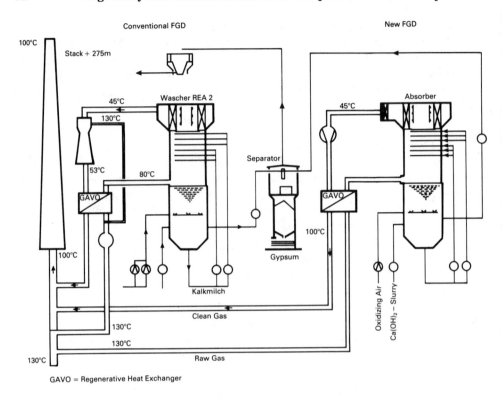

Fig. 4 — Flow Scheme of the conventional and the new flue gas desulphurisation plant of the 700 MW power station at Wilhelmshaven.

— to save energy for reheating by the use of regenerative heat exchangers.

Today, the efficiency of operating FGDs is over 90% and availability is comparable with other power station components. The quality of the produced gypsum is often better than the natural product. Estimations expect that the annual production of about 3.4×10^6 tonnes gypsum from 1995 onward can be absorbed by the market.

The Ordinance on Large Firing Installations prescribes a reheating of the cleaned gases up to at least 345 K in order to guarantee a better distribution of the flue gas in the atmosphere and to prevent liquid droplets from being emitted. This reheating is very energy-consuming if additional energy is supplied. Much energy can be saved by reheating the cleaned gas through regenerative heat exchange with the hot dirty gas. Appropriate devices using the Ljungstroem principle or similar are now state-of-the-art. The total additional energy demand of a wet lime/limestone FGD is estimated to be about 1% of the desulphurized electric capacity. The choice of lime or limestone is a matter of optimizing the local conditions. Special emission standards for FGD-wastewater are in preparation. Hazardous substances in the water like heavy metals, sulphates or fluorides are precipitated and filtrated; the remaining problem being the chlorides. A possible solution could be an evaporation of the water.

FGD processes other than wet lime/limestone scrubbing only play a subordinate part in the FRG. 30 820 MW$_{el}$ of the ordered FGD capacity of 33 160 MW$_{el}$ are based on lime or limestone. The ammonia-based Walther Process has 860 MW$_{el}$ commissioned (by-product: ammonia sulphate fertilizer). The activated charcoal process, sold by Uhde, has had only one order for a 230 MW$_{el}$ simultaneous SO$_2$/NO$_x$-reduction (by-product: SO$_2$-rich gas). Davy McKee recently received two orders for the desulphurization of the high sulphur lignite power stations at Buschhaus (350 MW$_{el}$) and Offleben (325 MW$_{el}$) (Wellman–Lord process, by-product: sulphur). Moreover, Davy McKee got the first order to desulphurize an industrial power station (700 MW$_{th}$, hard coal, by-product: liquid SO$_2$).

Spray-drying processes have produced only little interest. Fläkt received two orders (230 MW$_{el}$) and Niro Atomizer one (50 MW$_{el}$+110 MW$_{el}$).

3.2 Reduction of NO$_x$ emissions

Nitrogen oxide emissions from fossil fuel burning consists of more than 90% (average >95%) NO; the remaining emissions are NO$_2$. The origin of NO$_x$ emissions is twofold:

— fuel NO by conversion of chemically bound nitrogen in the fuel during the combustion to NO; and
— thermal NO by high temperature reactions of nitrogen and oxygen within the combustion air.

The usual nitrogen content of hard coals varies between 1 and 2% and residual oils contain about 0.3% nitrogen, whereas distillates and natural gas are nearly free of chemically bound nitrogen. Denitrification of fuels is complicated and uncommon; moreover it would be only partially effective. The main efforts are concentrated on the development of low NO$_x$ combustion technologies and high efficiency DeNO$_x$ flue gas treatments.

3.2.1 Combustion modification processes

The formation of thermal NO can be reduced by decreasing the temperature and the oxygen concentration in the firebox. The following measures are appropriate: a boiler design with low specific volumetric heat release to combustion systems with low temperatures; low excess air (LEA); low combustion air preheating; flue gas recirculation (FGR); low NO$_x$ burners (LNB); water/steam injection.

Fuel NO, important at coal firings, can be reduced by a stoichiometric staged combustion using, for instance, the overfire air technique (OFA). Special LNBs also exhibit a good reduction potential.

A recently developed low NO$_x$ combustion technology is the so-called in-furnace NO$_x$ reduction (IFNR) [15]. This process reduces already existing NO by second flame ('fuel staging') with a reducing atmosphere (local air deficiency). Fuel downstream of the residual air for total burnout is added. The various combustion modification (CM) measures can be combined, but the total reduction effect is smaller than the sum of the single effects; an overview is given in Table 8. All CM technologies can be easily retrofitted at existing facilities; the reduction effect, however, can differ as a result of site-specific constraints.

As can be seen from Tables 3 and 8, CM is not normally capable of meeting the

Table 8 — Reduction efficiency of low NO_x combustion modifications

Fuel type	Combustion system	Combustion modification	NO_x emission mg NO_2/m^3
Hard coal	Dry bottom pulverized coal firing	LNB	600–800
		LNB+OSC/FGR	400–650
		LNB+OSC+IFNR	200–400
	Wet bottom pulverized coal firing	LNB	1300–1800
		LNB+OSC+FGR+LEA	1000–1300
		LNB+OSC+FGR+LEA+IFNR	400–1000
	Fluidized bed combustion	atmospheric stationary FBC+OSC/FGR	300
		atmospheric circulating FBC+OSC/FGR	100–150
Rhinish lignite	Pulverized coal firing	FGR	400–600
		FGR+OSC+LNB	200–400
Oil		FGR	350–500
		FGR+OSC	250–400
		FGR+OSC+LNB	200–300
		FGR+OSC+LNB+IFNR	100–200
Gas		FGR	300–400
		FGR+OSC	200–300
		FGR+OSC+LNB	100–200
		FGR+OSC+LNB+IFNR	50–100

LNB, Low NO_x burner; LEA, Low excess air; OSC, Off stoichiometric (staged combustion); IFNR, in-furnace NO_x-reduction; FGR, flue gas recirculation.

stringent NO_x emission standards of the 13. BImSchV for facilities >300 MW_{th}, so that in any case post-combustion $DeNO_x$ devices must be installed. Nevertheless, there are many arguments in favour of both technologies. CM is much cheaper than flue gas denitrification. Therefore, for a fixed emission standard, every primary NO_x reduction allows the flue gas $DeNO_x$ device to be smaller and less expensive. Moreover, CM can be installed within a very short time and thus NO_x emission reductions are possible long before the installation of a $DeNO_x$ plant.

The NO_x emission standards for smaller and medium sized firing installations are oriented at primary NO_x reduction technologies. The draft proposal for an amendment of the TA Luft fixes the following NO_x limits: coal firings, 400 mg/m^3 (7% O_2, FBC=300 mg/m^3); oil firings, 300 mg/m^3 (3% O_2; distillates=250 mg/m^3); gas firings, 200 mg/m^3 (3% O_2). These values will be valid for new, licenced facilities (1–50 MW_{th}). The proposal provides a time period of five years for existing facilities to retrofit to the new standards, or a close down of the facility within ten years.

The present status of low NO_x combustion technology in the FRG can be characterized as follows: LNB or OFA have been installed in all new large coal firing installations since 1980 to meet an emission limit of 800 mg/m^3. Since 1984 a combination of LNB plus OFA has been introduced to meet an emission limit of 650 mg/m^3. Intensive investigations are now under way to reduce too high NO_x emissions of wet bottom coal firings down to 1000 mg/m^3 by combination of LEA, LNB, OFA, FGR and perhaps water injection.

A further drastic NO_x emission reduction is expected by a newly developed multiple stage mixing burner (MSM Burner, Fig. 5) [16]. This burner realizes the

Ch. 2.1] Regulatory and technical measures for air pollution control 35

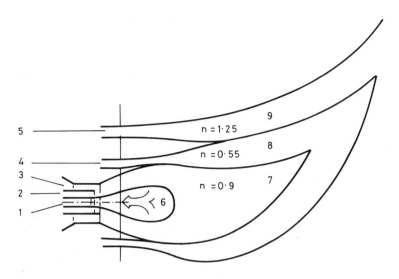

1 Primary Air
2 Fuel 1 and Transport Air
3 Secondary Air (swirled)
4 Fuel 2 and Transport Air
5 Staged Air
6 Internal Recirculation Zone
7 Primary Flame
8 Secondary Flame
9 Burn-out Zone

Fig. 5 — Multiple staged mixing burner.

IFNR principle by fuel staging at the burner (and not, as is usual, in the firebox). In laboratory tests with hard coal NO_x levels of 300–400 mg/m³ were achieved; using Rhinish Lignite even lower NO_x concentrations were possible. The same burner type has now been tested with great success at a 125 MW_{th} gas boiler. NO_x base emissions of >300 mg/m³ could be reduced to <100 mg/m³ (in combination with external FGR).

For oil firings, too, a new LNB has been developed which combines the principles of air staging and flue gas recirculation [17]. At a 110 T/h steam boiler (residual fuel oil) base emissions of 540 mg/m³ have been reduced down to 350 mg/m³.

A very promising low NO_x combustion technology is represented by fluidized bed combustion (FBC). Existing stationary FBCs with NO_x levels in the range 300–600 mg/m³ will be reduced by air staging down to <300 mg/m³. The NO_x emission level of the above mentioned 84-MW circulating FBC is already below 200 mg/m³ [13].

All retrofitted CM technologies for facilities >50 MW will be installed by July 1988.

3.2.2 Flue gas denitrification

The stringent NO_x emission standards for firing installations >300 MW_{th} can only be met by flue gas denitrification technologies. Since the decision of the UMK in April,

1984, a lot of activity has been directed towards transferring Japanese experience with DeNO$_x$ plants to site-specific German conditions. This concerns in particular:

— operating conditions (Japan: base load, few starts, FRG middle load, frequent starts);
— different coal composition (possible negative influence on catalyst);
— different combustion systems (the German slag tap firing is unusual in Japan); and
— different regulations (safety rules, quality standards for fly ash, waste water, etc.).

In 1985 about fifty small DeNO$_x$ pilot plants will be operating at various types of German power plants. The main objective is not only transfer and optimization of Japanese technologies but also local development.

Table 9 gives an overview on the various DeNO$_x$ techniques in the FRG. The

Table 9 — Flue gas treatment techniques for NO$_x$ and NO$_x$/SO$_2$ removal in the Federal Republic of Germany

Vendor/Licensee/Developer	Process	System
Deutsche Babcock Anlagen AG, Krefeld	Kawasaki Heavy Industries (KHI)	Selective catalytic reduction (SCR)
Energie- und verfahrenstechnik GmbH (EVT), Stuttgart	Mitsubishi Heavy Industries (MHI)	Selective catalytic reduction (SCR)
Kernforschungszentrum Karlsruhe und universitat Karlsruhe		Electron beam process for simultaneous SO$_2$/NO$_x$ removal (addition of NO$_3$; product ammonium sulphate and nitrate)
Lentjes, Dusseldorf und Gottfried Bischoff GmbH & Co. KG, Essen	Lentjes	3-way catalyst for NO$_x$ reduction
Niro Atomizer Trocknungsanlagen GmbH, Karlsruhe	Niro Atomizer	Extension of dry absorption process (addition of NaOH)
Saarberg-Holter-Lurgi GmbH (SHL) Saarbrucken and Universitat Essen[a]		Extension of lime-limestone scrubbing process (addition of EDTA)
L&C Steinmuller GmbH, Gummersbach	Ishikawajima Harima Heavy Industries (IHI) (Cooperation)	Selective catalytic reduction (SCR)
Steuler-Industriewerke GmbH Hohr-Grenzhausen	Steuler	Selective catalytic reduction (SCR)
Thyssen Engineering GmbH Essen	Mitsubishi Heavy Industrie (MHI)	Selective catalytic reduction (SCR)
Uhde GmbH Dortmund	Bergbau-Forschung/Uhde	Simultaneous SO$_3$/NO$_x$ flue gas cleaning with activated coke and NH$_3$
	Babcock-Hitachi KK (BHK)	Selective catalytic reduction (SCR)
Universitat Essen[a] (Saarbergwerke Saarbrucken and Steag Essen)		Selective non-catalytic reduction (SNR)
Walther & Cie, AG Koln	Walther	Extension of Walther FGD process for simultaneous SO$_2$/NO$_x$-removal (NH$_3$ scrubbing product ammonium-sulphate and nitrate)

[a] Research and development.

predominating technique offered is selective catalytic reduction (SCR). This works as follows: at temperatures between 300 and 400°C, NH_3 is injected into the flue gas. Downstream of the injection the flue gas passes a special catalytic reactor where NH_3 reacts with NO_x forming nitrogen and water vapour. Usual NO_x emission reduction efficiencies are about 80%. The different realizations of the process vary with respect to catalyst compositions, geometry and location. The most common SCR plants consist of ceramic honeycomb or plate-type catalysts which are installed downstream of the economizer and upstream of the air preheater in the high dust-containing flue gas (high dust systems). Low dust systems use a high temperature electrostatic precipitator to protect the catalyst from dust erosion and dust corrosion. Another variant locates the SCR plant after the FGD in the cooled clean gas. This system needs additional reheating energy but it has the advantage of being easily retrofitted with minimal space problems and is independent of coal compositions and combustion systems. There are also no problems with possible NH_3 leakage.

The most important problems at SCR plants that should be considered are:

— minimizing the NH_3 leakage (<5 p.p.m.);
— protecting the air preheater and other downstream devices against depositions (ammonia sulphate, ammonia bisulphate);
— ammonia absorption by the fly ash; and
— pollution of the FGD wastewater by ammonia.

These problems, however, are not a major barrier to the application of SCR technology. There is only a need for further technical optimization, and flue gas denitrification at large firing installations is regarded as state-of-the-art.

The first commercial large scale SCR installation is now under construction at the hard-coal fired (dry bottom) 460 MW_{el} power station at Altbach (operator: Neckarwerke AG). It is scheduled to go into operation by the end of 1985. A high dust system is used to reduce NO_x emissions to below 200 mg/m^3. The first SCR plant for a wet bottom coal firing will start operation at the 345 MW_{el} power station at Knepper (operator: VKR) in early 1986. Here, too, a high dust system will be installed. Several other SCR plants for coal-fired power plants are in the planning phase.

Further developments are directed towards finding low temperature catalysts and improving the regulation of the NH_3 injection for lower NH_3 leakage. Other systems like the activated charcoal process and the simultaneous SO_2/NO_x scrubbing by the Walther process will be also realized in commercial plants. It can be expected that in the early 1990s power stations with a capacity of more than 50000 MW_{el} will be equipped with flue gas $DeNO_x$ devices.

3.3 Reduction of dust emissions

In the FRG all coal fired utilities are equipped with electrostatic precipitators (ESP) for a reduction of dust emissions. In 1974 the TA Luft introduced an emission limit of 150 mg/m^3 for watertube boilers (other smaller boilers: 300 mg/m^3) with the additional requirement that the failure of one section of the ESP must not result in the emission limit being exceeded. This led to ESPs with three or four sections and a specific collection area of 100–200 m^2/m^3/s; the dust emissions range from 20 to 50 mg/m^3 under normal operating conditions. For lignite firings two or three sections with a specific collection area of about 50 m^2/m^3/s are sufficient to achieve emission concentrations of 25 to 90 mg/m^3 because of the more favourable dust characteristics.

For small- and medium-sized industrial coal firings, cyclons have been used as dust separators. By the end of the 1970s some fabric filters were installed at these facilities. This development is now increasing and accelerating because of the new emission standard of 50 mg/m^3 proposed in the draft of the TA Luft amendment for coal and wood firings with heat inputs in the range 1–50 MW$_{th}$. This emission limit cannot be met by cyclons and therefore ESPs or fabric filters are necessary.

At present, over thirty fabric filters are operating at coal firings, the largest at a 183 MW$_{th}$ wet bottom boiler, the smallest at a 0.5 MW$_{th}$ fluidized bed combustion system. Usually the dust concentration in the flue gas is far below 20 mg/m^3. Both bag cleaning systems, reverse flow — preferable for larger units — and pulse jet, are installed, the filter rate ranging from 0.6 to 1.0 m^3/m^2/min. It is expected that the increasing installation of direct desulphurization by sorbent addition at small- and medium-sized firing installations will be accompanied by the installation of fabric filters.

Only a few ESPs are installed at oil firings. In most cases the quality of the fuel and of the combustion system are sufficient to meet an emission limit of 50 mg/m^3. The installation of fabric filters is also possible as was demonstrated by a 3000-m^3/h pilot plant. Gas firings — converter firings in the steel industry exempted — are operating without any dust separators.

The aim of stringent dust emission limits is not only to reduce the total dust emission but also to reduce the emission of hazardous and toxic substances like heavy metals. For normal fuels it is supposed that an emission limit of 50 mg/m^3 simultaneously reduces the emission of these substances to tolerable limits. Additional measures are required if there are indications that the fuel is highly contaminated with heavy metals. In these cases the 13. BImSchV limits the emission of As, Pb, Cd, Cr, Co and Ni and their compounds to 0.5 mg/m^3 (oil: 2 mg/m^3).

There are many possibilities for retrofitting existing facilities: cyclons must be replaced by ESPs or fabric filters. If existing ESPs only marginally exceed the emission limit a further homogenization of the gas stream and some improvement of the high voltage supply may be sufficient. If there are separation problems with dust having a high electric resistance, SO$_3$ doping of the flue gas can result in an improvement. In other cases the ESP should normally be replaced or extended by a further section; in some cases a parallel fabric filter can also be a solution.

Arbitrary estimations of the investements for ESPs and fabric filters are given in Figs 6 and 7 [18]. There are no clear cost advantages between the two systems, and therefore site-specific considerations are necessary in every case to choose between ESP and fabric filter.

According to the 13. BImSchV, all measures for an additional dust emission reduction must be accomplished by July 1988.

3.4 Mobile sources

Technical details of emission reduction technologies for mobile sources are not dealt with in this chapter. In the view of the FRG no known advanced combustion techniques are sufficient to achieve the necessary emission reduction. The only

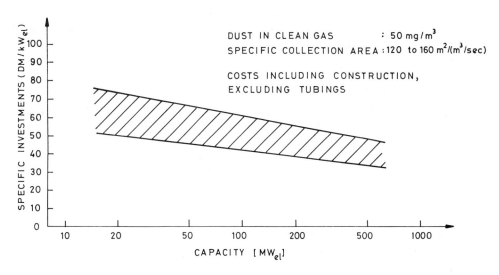

Fig. 6 — Specific investments for electrostatic precipitators at coal firings.

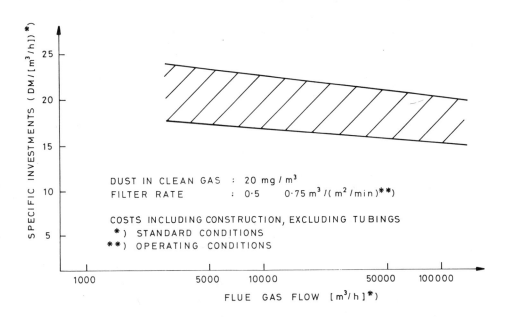

Fig. 7 — Specific investments for fabric filters at coal firings.

acceptable way to achieve a reasonable emission reduction is the use of a three-way catalyst in combination with a sensible air/flue mixture control. These systems provide NO_x control of 90–95% and have been in production in the USA for eight years.

The catalytic converter is poisoned by lead, and therefore unleaded petrol must be supplied. Since the end of 1984 the mineral oil industry in the FRG has supplied the filling stations with unleaded petrol and an increasing number of motor vehicles fitted with catalytic converters are on German roads. The federal government intends to introduce financial incentives to expedite the transition to low pollution vehicles (tax reductions for unleaded petrol, tax exemptions for low pollution vehicles).

4. SUMMARY

Damage to forests, waters and buildings in the FRG and in other European countries is increasing at an alarming rate. Available knowledge indicates that airborne pollutants which have polluted the air since the beginning of industrialization, and are still increasing, are a decisive factor in this process. Therefore increased efforts of all states are necessary to reduce pollution levels in the air. Numerous suitable methods are available in the field of energy conservation, the use of low pollution fuels, improved combustion processes and waste gas purification.

In the FRG all possible regulative and technical measures have been taken to bring about rapid realization of an advanced state-of-the-art at all new and existing air pollution sources. National efforts alone are not sufficient, however, to combat transboundary air pollution effectively. Therefore the FRG supports all international cooperation activities in the field of air pollution control.

REFERENCES

[1] *Bundes-Immissionsschutzgesetz* (BImSchG), as of 15.3.1974, BGB1, I.S. 721; Last amendment 4.3.1982, BGB1, I.S. 281.
[2] *Verordnung über Genehmigungsbedürftige Anlagen* (4. BImSchV) as of 14.2.1975, BGB1. I.S. 499, 727; amended 27.6.1980, BGB1, I. S. 772.
[3] *Verordnung über Schwefelgehalt von Leichtem Heizöl und Diesel- kraftstoff* (3. BImSchV) as of 15.1.1975, BGB1. I.S. 264.
[4] *Verordnung über Feuerungsanlagen* (1. BImSchV) as of 5.2.1979, BGB1. I.S. 165.
[5] *Verordnung über Großfeuerungsanlagen* (13. BImSchV) as of 22.6.1983, BGB1. I.S. 719.
[6] *Technische Anleitung Zur Reinhaltung der Luft* (TA Luft) as of 28.8.1974, GMB1. S. 426, amended 23.2.1983, GMB1 S. 94.
[7] Schärer, B.; Haug, N. On the economics of the flue gas desulphurization. In *Cost of Coal Pollution Abatement,* E.S. Rubin and I.M. Torrens (eds), OECD, 1983, pp 133/143.
[8] *Wärmeschutzverordnung* 24.2.1982, BGBl, I.S. 209.
[9] *Heizungsanlagen-Verordnung* (HeizAnlV) as of 24.2.1982, BGB1. I. S.205.
[10] Commission of the EEC. *Proposal of the Council for a Guideline on Air*

Pollution Control of Large Firing Installations. Doc. COM (83) 704, as of 13.12.1983.

[11] Hein, R.G. and Kirchen, G. *Reduction of SO_2 Emissions in Brown Coal Combustion: Results from Research and Large Scale Demonstration Plants* 1st Joint Symposium on Dry SO_2 and Simultaneous SO_2/NO_x Control Technologies, EPA/EPRI November 13–16, 1984, San Diego, CA.

[12] Wagenknecht, P. and Puder, F. *Emissionen von und Betriebserfahrungen mit Wirbelschichtfeuerungen* Informationen zur Raumentwicklung (1984) Nr. 7/8, S. 811/826.

[13] Schmitz, W. and Köhnk, P. *Verbrennung von Ballastkohle in der Zirkulierenden Wirbelschicht, Erste Erfahrungen an Einer Industrie Anlage zur Versorgung Eines Chemischen Werkes mit Prozeßwarme und Strom* VDI Bericht Nr 511 (1984) S. 153/158.

[14] Chughtai, Y., Michelfelder, S. and Leikert, K. *Operation and Performance Report of the Steinmuller Low NO_x SM Burner and its Potential Towards Utility Boiler SO_x Control via Sorbent Injection.* Proceedings of the 1982 Joint Symposium on Stationary Combustion NO_x Control, EPRI CS-3182, July 1983.

[15] Takahashi, Y., Sakai, M., Kunimoto, T., Ohme, S., Haneda, H., Kawamura, T. and Kaneko, S. *Development of MACT In-Furnace NO^x Removal Process for Steam Generators.* Proceedings of the 1982 Joint Symposium on Stationary Combustion NO_x Control, Vol. 1, EPRI CS-3181, S. 15-1/20.

[16] Leikert, K. *Feuerungstechnische Massnahmen zur Senkung der NO_x-Emission* Steinmuller-Kolloquium, Gummersbach, Marz 1984.

[17] Information Paper from Deutsche Babcock AG, Oberhausen, 1984.

[18] Davids, P. and Lange, M. *Die Grossfeuerungsanlagen-Verordnung. Technischer Kommentar, VDI Verlag, GmbH, Dusseldorf, 1984.*

2.2

Abatement technologies for air pollutants

J. J. Stukel
Director, Engineering Experiment Station, and Associate Dean, Professor of Environmental Engineering and Mechanical Engineering, College of Engineering, University of Illinois at Urbana-Champaign, Urbana, Illinois, USA

1. INTRODUCTION

Emissions from the traditional sources are well known. The visible nature of particulate emissions cause plant shutdowns and reduce operations in order to ensure that plants meet particulate emissions standards. The costs of the reductions in operation capacity often result in high costs of compliance and the need to design systems to ensure that the standards are met. The particulate costs however, are industry, emission limit and process specific. For example, for most industrial processes the capital cost of meeting current new source performance standards (NSPS) range from $5 to $50 (1983$) per actual cubic feet per minute of gas treated. For most non-combustion sources, this cost is the major (often only) air pollution control expenditure. For coal combustion, the primary technologies used to control particulate emissions are electrostatic precipitators (ESP), baghouses and scrubbers. Typical operating and capital costs for a 500 MW boiler are shown in Fig. 1. Both the capital and operating costs of particulate control for ESPs depend on the alternative methods used to control sulphur oxide emissions as shown in Fig. 1. This shows that control of sulphur oxide emissions by use of low sulphur coal to meet the NSPS can increase the cost of particulate control utilizing an ESP by a factor of at least 1.5. The high costs of particulate control, especially for non-combustion sources, led to political pressure to relax the particulate standards. To gain some idea as to the magnitude of the problem, it should be noted that about 300 of the air quality control regions in the United States are currently not in compliance with the particulate and air quality standards. The costs associated with bringing these sources into compliance would be very large.

Any attempt to review emerging technologies for the control of air pollutants in the United States is going to be incomplete. There are a myriad of private, industrial

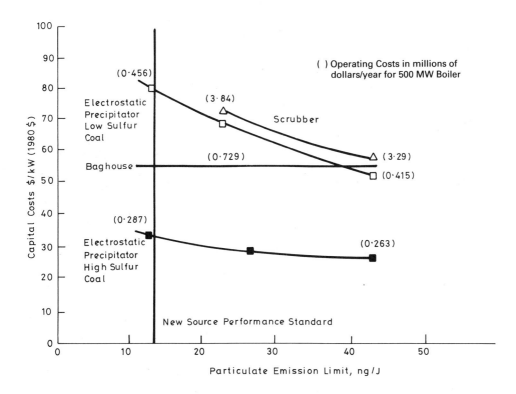

Fig. 1 — Cost of particulate matter control in a coal utility boiler [1].

and government research groups currently working on new methods of control for any number of toxic materials. This conference, however, is focusing on control technologies for existing and potential air pollutants. Since the Environmental Protection Agency (EPA) is responsible for setting air quality and emission standards in the United States, a review of their research and development programme would seem appropriate. Therefore, much of the material presented in this chapter is taken from EPA briefing materials and private communication with EPA researchers. The author gratefully acknowledges their input.

The chapter first addresses the control of particulate emissions followed by control of gaseous emissions. In all cases, the discussion touches briefly on existing technologies then tries to present emerging control technologies for a number of pollutants.

2. CONTROL OF PARTICULATE EMISSIONS

The particulate control regulatory programme in the United States seeks to develop effective cost efficient particulate control methods to control combustion and industrial source emissions, develop fugitive emission control methods for particulate sources, and characterize particulate emissions. The particulate regulatory programme is basically broken down into two parts. One part deals with the traditional sources and the other deals with the non-traditional sources.

2.1 Control of emissions from traditional sources

Particulate emissions from non-traditional sources are very large. The non-traditional sources are usually labelled fugitive sources. The fine fraction of a typical urban dust is dominated by uncontrolled stack emissions, secondary particles and volatile or moderately volatile elements from combustion sources. Typically, the coarse fraction is dominated by particles from fugitive sources. Another important point is that fugitive emissions, which are relatively uncontrolled, have a significant contribution in the PM10 size range. Fugitive emissions contribute significantly to the approximately 300 air quality regions mentioned above that are currently in non-attainment.

2.2 Emerging control methods for traditional sources

The objectives of the EPA particulate control research programme in the United States are aimed at developing breakthrough ESP and fabric filter technology, and innovative operation and maintenance procedures. The programme also seeks to prepare inhalable particulate emission factors and to find ways to reduce fugitive and visible emissions.

The particulate control programme in the United States has been vigorous. The programme has developed:

(1) A concept of enhancing fabric filtration with electrostatic fields which have the potential for a 50% cost reduction without loss of efficiency.
(2) Two- and multi-stage ESPs which offer the potential for 50% or greater cost reduction for the collection of particles from low sulphur coal burning facilities.
(3) Large diameter electrodes for ESPs that can greatly reduce particulate control costs in both new plants and for retrofit applications.
(4) Performance and cost models for particulate control technology that allow rapid, accurate estimation of the performance and cost of new installations and can be used for diagnosing poorly performing existing units.
(5) Flue gas sodium conditioning technology.
(6) New technologies for the control of fugitive emissions.

I would like now to discuss briefly each of these new emerging abatement control technology options for particulates.

In an electrostatically enhanced fabric filter (ESFF) an electric field of 2–4 kV/cm parallel to the fabric filtration surface is developed between electrodes woven into the fabric or placed in close proximity to it. Natural charges upon the particulate matter cause them to be preferentially attracted to the fabric on or near the electrodes. The uneven deposition of dust upon the fabric provides areas of higher porosity or lower pressure drop. This allows the choice of operating the baghouse at a lower pressure drop for a savings in operating cost or at a higher flow rate for capital investment savings. Increasing the charge of the particles, by precharging, allows still further performance improvement.

EPA-sponsored projects have operated the ESFF successfully in the reversed air mode on pilot plants using coal utility sized bags. Projects have also operated the ESFF concept successfully in pulse jet modes in pilot plants on an industrial blower. The results demonstrate that there is a 50% reduction in pressure drop for both the reverse air and pulse jet modes of operation with a resultant decrease in operating or capital costs. EPA tests have also utilized bags with woven-in electrodes and have shown that these will operate without significant problems. Currently, parametric design data are being developed in the EPA laboratories for various kinds of fly ash and operating conditions. Even more spectacular reductions in pressure drop have been achieved in the laboratory by precharging the aerosols to be collected. Figure 2

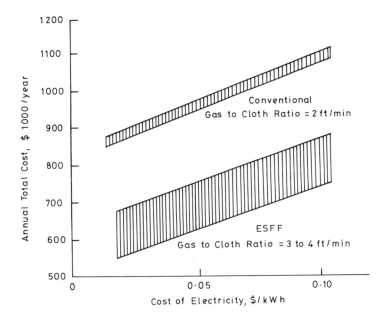

Fig. 2 — Cost comparison for a reverse-air baghouse in a 100 MW new power plant [1].

shows the cost comparison for a reverse air baghouse for a 100-MW new power plant compared to the conventional baghouse operation. As can be seen there is a significant reduction in the cost of operation when operating under the ESFF mode. Likewise, examination of Fig. 3, which gives the cost comparison for a pulse jet baghouse operating at 40 000 actual cubic feet per minute, shows again that there is a significant (about one-half) reduction in the total annual operating cost when using the ESFF concept.

Conventional ESPs operate by first charging the particle and then removing it from gas streams by electrostatic attraction. In these types of configurations, charging and removal of the particle are done simultaneously. This method works well enough for low resistivity fly ash (typically from high sulphur coal), but does not

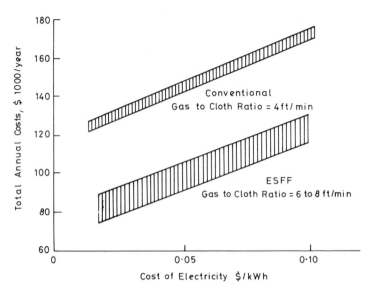

Fig. 3 — Cost comparison for a pulse-jet baghouse at 40 000 actual cubic feet per minute [1].

work very well with high resistivity fly ash obtained from the burning of low sulphur coal. Because of this difference, it is necessary to build much larger ESPs at a considerably greater cost to achieve the same efficiency for high resistivity as for low resistivity fly ash. By separating the charging and collection function into a precharger and collector stage, each operation can be optimized. The application of this concept results in an ESP which is about half the size of a conventional ESP for high resistivity dusts. An EPA-sponsored project is operating a two-stage ESP successfully on a 10-MW (30 000 ACFM) pilot unit at the TVA Bull Run Utility, burning low sulphur coal having high resistivity ash. This unit achieved the same efficiency as a conventional single-stage ESP having twice the plate electrode area for the same flow rate. The retrofitability of this technology to a conventional ESP was shown to be feasible on this pilot unit. Based on this test, additional tests were run using the two-stage ESP concept on the in-house ESP. The results of this test indicated that the two-stage ESP operated very well on several different fly ashes available to EPA. The three types of prechargers which are currently being developed are the tri-electrode, the cold-pipe and the charged droplet precharges. Figure 4 gives a comparison of the installed cost of the conventional ESP versus the two-stage collection for high resistivity fly ash. As can be seen, the installed cost of the conventional ESP escalated quite rapidly with increase in fly ash resistivity. In contrast, the two-stage ESP collection characteristics remain unchanged as the fly ash resistivity increases. This results in very large cost savings.

Two-stage ESPs in which the charging and collection functions are separated can be improved still further by operating the unit with several precharger stages each followed by its own collector. This multi-stage approach results in an ESP, most applicable to new installations, which can achieve NSPS levels of control with about

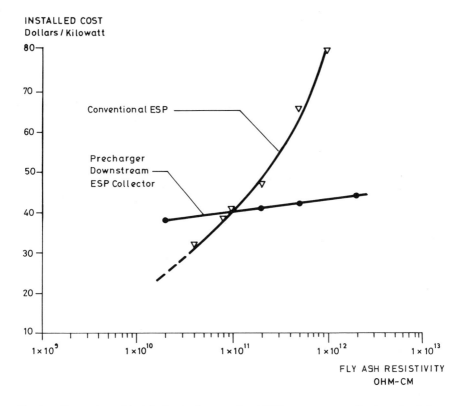

Fig. 4 — Comparison of installed cost of conventional ESP vs. two-stage collection of resistivity fly ash [1].

one-fifth the electrode area of a conventional single-stage ESP, both operating on high resistivity fly ash from the combustion of low sulphur coal. This concept was successfully tested on a 3 MW (10 000 ACFM) pilot ESP where it achieved NSPS levels of control with an electrode plate area one-fifth of that needed for a conventional single-stage ESP when operating with a high resistivity fly ash from the burning of low sulphur Western coal. The results from this experiment verified EPA results obtained from a smaller in-house ESP pilot unit operating with other high resistivity fly ash. Figure 5 gives a comparison of typical single- and two-stage ESP operations on high resistivity fly ash. Examination of the figure reveals that one can get very high mass collection efficiencies with a very low specific collection area using the two-stage ESP with the cooled electrode precharger.

Historically, conventional single-stage ESPs were designed and built with smaller diameter (1/8 inch) wire discharge electrodes which operated well when controlling low resistivity fly ash from the burning of high sulphur coal. When ESPs were designed to control high resistivity ash from the burning of low sulphur coal the practice of using small diameter electrodes continued. However, an in-house EPA programme showed that by using large diameter (3/8 inch) wire discharge electrodes,

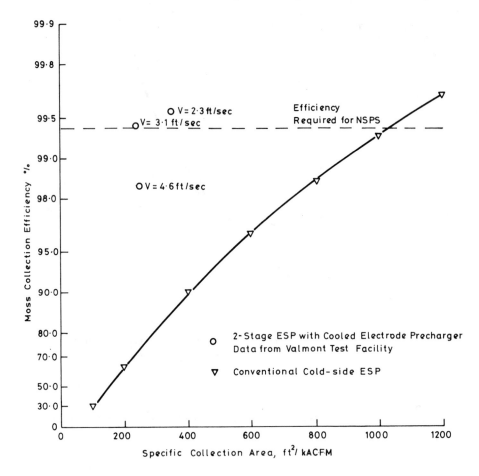

Fig. 5 — Comparison of typical single- and two-stage ESP operations on high resistivity ($\simeq 10^{12}$ Ω/cm) fly ash [1].

reductions of emissions by a factor of four were achievable with high resistivity fly ash as compared to small diameter electrodes. With 1/8-inch wires the current starts to flow uncontrollably almost immediately upon corona onset, with an immediate decrease in voltage which stops the ESP from operating. With large diameter electrodes, the current starts to flow at the corona onset point and there is a range where it varies with increasing voltage. This provides an operating range, before control of the current is lost, in which both charging and collection can occur. This technology is useful for retrofit applications at a very low cost of $3–$5 (1983$)/kW as compared to a replacement cost of $70–80/kW. The results of some pilot tests are shown in Table 1, showing that the addition of a large diameter electrode to the 10 MW (10 000 actual cubic feet per minute) Bull Run Pilot unit, which had a high resistivity fly ash, resulted in a factor of four reduction of emissions. These results were verified on the EPA in-house ESP pilot unit operating with other high or moderate resistivity fly ashes.

Table 1 — Difference in ESP performance (Bull Run Data)

	3/8-inch wires	1/8-inch wires
Electric field strength (kV/cm)	4	3
Current density ($\mu A/ft^2$)	4.7	6.8
Collection efficiency (%)	99.64	98.58
Penetration	0.0036	0.0142

In the early 1970s utilities began to burn large amounts of low sulphur coal. This led to problems with performance and design certainty in conventional cold side (300°F) ESP. Because short-term empirical data showed that the electrical resistivity of fly ash decreased at high temperature (650-800°F), most ESP vendors and utilities decided to place the ESP before the air heater (hot-side) to take advantage of the lower resistivity. Thus over 2000 MW of electric generating capacity were installed with hot-side ESPs. All units met standards upon start-up. However, over half of the units began to experience performance problems within a few months. In most cases, the problems resulted in failure to meet particulate emission standards under full load operations. These plants were then forced to operate at reduced load and/or shut down every month or so to clean the ESP.

Fundamental research in EPA laboratories showed that the resistivity of fly ash at hot-side temperatures depends on the concentration of sodium ions in the ash. (The electric current is carried by migration of sodium ions.) Under the conditions of hot-side ESP operation, the sodium ions are removed from the fly ash near the collector. This thin, low sodium concentration layer is the primary cause of the loss of performance in the hot-side ESP. The research also showed that the problem could be overcome by adding small amounts of sodium carbonate or sulphate, or any other sodium salt, to the coal. A full-scale demonstration of sodium conditioning was co-funded by EPA, EPRI, Southern Company Services, and Belt Power Lancing Smith Station with complete success. Emissions were reduced from over 0.35 lb/MBtu to less than 0.05 lb/MBtu at a cost of 25¢ per ton of coal. Cost savings were about $1 000 000 a year. Sodium conditioning has been adopted as a permanent solution.

The main factor that limits the use of sodium conditioning is the uncertainty surrounding the effects of adding sodium on the long-term operation of the boiler. It is well known that high sodium ashes cause fouling and corrosion in boilers, and consequently there is concern that adding sodium for improving ESP operation will cause boiler problems. There was no evidence of such problems at Lancing Smith.

The sodium conditioning method has been demonstrated at full scale as a method for reduction of emissions from hot-side ESPs. In addition, design methods have been developed to enable prediction of the effects of sodium addition and to determine the level of sodium conditioning needed. Research co-funded by EPA and Southern Company Service is underway to determine the effects of sodium conditioning on boiler operations.

2.3 Control methods for non-traditional sources

Fugitive (non-traditional) particulate emissions represent the largest contributor to ambient particle levels in the United States, in many instances five to eight times the contribution of particulates from ducted (traditional) sources. Therefore, by developing and applying even moderately effective control strategies to fugitive emissions, a significant improvement in ambient air quality can result. However, the sources of fugitive emissions are diverse, often covering a large area and are in general difficult to control. This necessitates the development of innovative and highly specialized control technology.

The control technologies under investigation for the control of fugitive particulate emissions include an improved street sweeper to reduce fugitive emissions from paved roads, and chemical dust suppressants and road carpets for the control of emissions from unpaved roads.

Field tests by EPA have established the initial control efficiencies realized by sweeping and flushing paved roads with water and the application of Petrotac and Coherex for unpaved roads. The life cycle controlling efficiency measurements for Petrotac have been completed on one unpaved road segment. For Coherex, the life cycle controlling efficiency measurements on one unpaved road segment over an initial cycle and a first reapplication cycle have also been completed. The results of the Coherex tests shown in Fig. 6, indicate that the emissions from the road after an

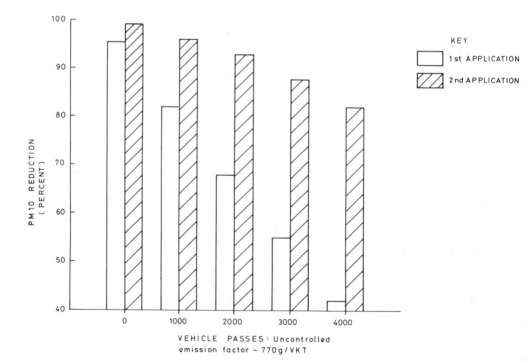

Fig. 6 — Coherex dust suppression effectiveness on PM10 emissions [1].

initial application were only 6.8% greater than the allowable emissions. However, after 4000 vehicle passes the emissions were 13.7% above the limit. After a second application of Coherex, initial emissions were in excess of the allowable limit by 5% whereas after 4000 vehicle passes the excess had risen to 17.8%.

Additional research has been conducted to evaluate the effectiveness of wind screens as a method of controlling emissions from storage piles. Charged fog has been demonstrated as an effective technique on selected industrial processes. Finally, air curtain technology is being evaluated as a means to reduce industrial emissions, particularly when buoyant plumes are present.

3. CONTROL OF GASEOUS EMISSIONS

The gaseous control regulations programme in the United States is focusing on developing control methods for sulphur oxides, oxides of nitrogen and volatile organic compounds (VOC).

3.1 Control of sulphur oxide emissions

3.1.1 Current control methods

Perhaps the most visible area of environmental research in recent years in the United States has been methods to control sulphur oxide emissions from electric utility boilers. Today there are many utility installations which have operating sulphur oxide control equipment. Examination of Table 2 reveals that there are 116

Table 2 — Number and total capacity of FGD systems [2]

Status	No. of units	Total controlled capacity^a (MW)	Equivalent scrubbed capacity^b (MW)
Operational	116	46 801	43 186
Under construction	26	14 727	14 622
Planned:			
Contract awarded	21	12 909	12 635
Letter of intent	7	6 060	6 060
Requesting/evaluating bids	3	1 775	1 775
Considering only FGD systems for SO_2 control	41	23 878	23 700
Total	214	106 150	101 978

^{a}Summation of the gross unit capacities (MW) brought into compliance by the use of FGD systems, regardless of the percentage of the flue gas scrubbed by the FGD system(s).
^{b}Summation of the effective scrubbed flue gas capacities in equivalent MW, based on the percentage of flue gas scrubbed by the FGD system(s).

operational units consisting of 46 801 MW of controlled capacity. The equivalent controlled scrubbed capacity for these units is 43 186 MW. Table 2 also shows that 26 units (14 727 MW) are currently under construction. The total number either in operation, under construction or planned, amounts to 214 units with a controlled capacity of 106 105 MW (September 1983). Table 3 gives information on the types of

Table 3 — Summary of FGD systems by process (percentage of total MW) [2]

Process	By-product	Sept. 1983	Dec. 1999	Dec. 1999 (Normalized)[a]
Throwaway-product process				
Wet systems				
Lime		25.0	13.8	17.3
Limestone		46.2	40.8	51.2
Lime/				
alkaline fly ash		3.5	1.5	1.9
Limestone/				
alkaline fly ash		6.6	4.3	5.4
Dual alkali		3.6	2.2	2.7
Sodium carbonate		3.5	3.1	3.9
NA[b]		—	4.0	5.0
Dry systems				
Lime		3.6	6.2	7.8
Sodium carbonate		1.0	0.4	0.5
Saleable-product process				
Aqueous carbonate/	Elemental			
spray drying	sulphur	0.2	0.1	0.1
Lime	Gypsum	0.2	0.1	0.1
Limestone	Gypsum	0.4	0.6	0.8
Magnesium oxide	Sulphuric acid	1.7	0.7	0.9
Wellman Lord	Sulphuric acid	4.5	1.9	2.4
Process undecided		—	20.3	—
Total		100.0	100.0	100.0

[a] The effect of those systems listed as 'Process undecided' is removed.
[b] NA, not available (these systems are committed to a throwaway-product process; however, the actual process is unknown at this time).

FGD systems in use or planned. Lime and limestone systems are installed on 71.2% of the total MW being controlled in September 1983. This trend is expected to continue into the 1990s. Table 4 gives the reported and adjusted capital and annual

Table 4 — Categorical results of the reported and adjusted capital and annual costs for operational FGD systems [2]

	Reported						Adjusted					
	Capital			Annual			Capital			Annual		
	Range ($/kW)	Average ($/kW)	σ	Range (mills/kW/h)	Average (mills/kW/h)	σ	Range ($/kW)	Average ($/kW)	σ	Range (mills/kW/h)	Average (mills/kW/h)	σ
All	23.7–213.6	80.2	44.3	0.1–13.0	2.3	2.8	38.3–282.2	118.8	58.1	1.6–20.8	7.6	4.1
New	23.7–213.6	80.4	46.1	0.1–5.5	1.7	1.8	38.3–263.9	110.8	48.4	1.6–14.6	6.8	3.2
Retrofit	29.4–157.4	79.7	39.4	0.5–13.0	4.5	4.4	60.4–282.2	139.3	73.8	4.3–20.8	9.7	5.3
Saleable	132.8–185.0	153.1	20.6	13.0–13.0	13.0	0.0	254.6–282.2	271.6	12.1	16.7–20.8	18.1	1.9
Throwaway	23.7–213.6	75.8	41.5	0.1–11.3	2.1	2.4	38.3–263.9	110.9	47.6	1.6–17.6	7.0	3.4
Alkaline fly ash/lime	43.4–173.8	93.9	44.0	0.4–5.4	2.1	1.9	52.5–184.4	122.8	51.4	3.0–14.1	7.2	3.8
Alkaline fly ash/limestone	49.3–49.3	49.3	0.0	0.8–0.8	0.8	0.0	102.6–102.6	102.6	0.0	5.4–5.4	5.4	0.0
Dual alkali	47.2–174.8	97.8	55.3	1.3–1.3	1.3	0.0	87.8–263.9	146.7	82.9	5.0–13.9	8.7	3.8
Lime	29.4–213.6	81.8	43.7	0.3–11.3	3.2	2.7	60.4–210.0	116.5	44.2	4.0–17.6	8.1	3.6
Limestone	23.7–170.4	67.9	37.2	0.1–7.8	1.6	2.2	38.3–194.3	98.9	44.0	1.6–14.6	6.1	3.1
Sodium carbonate	42.9–100.8	69.2	26.6	0.2–0.5	0.4	0.1	87.1–150.9	110.9	26.4	5.8–7.4	6.4	0.7
Wellman Lord	132.8–185.0	153.1	20.6	13.0–13.0	13.0	0.0	254.6–282.2	271.6	12.1	16.7–20.8	18.1	1.9

costs for operational flue gas desulphurization (FGD) systems (September 1983), and shows that both the capital and the operating costs are quite variable, probably due to site conditions. Lime and limestone units are the cheapest options. Sulphur oxide controls are also an important element in the control of fine particles and of acid deposition, in addition to being important for the maintenance of national ambient air quality control standards for SO_2. Today's control technology is applied primarily to large new sources of sulphur oxides but control of other pollution sources such as industrial boilers and existing utility boiler sources may be necessary. Therefore, further research and development on control of sulphur oxides can have substantial impact, particularly in the areas of improving the performance, reliability and cost effectiveness of existing control technologies.

3.1.2 Emerging control methods .

While substantial sulphur oxides control research has been undertaken in the past, further research and development on oxides and sulphur control can have substantial benefit. Maintenance of SO_2 standards, acid deposition and future growth in coal use give rise to the need for development of increasingly cost effective control technology. High costs for control have constrained application of control technology, especially for the retrofit situation. Because of this, research on application of lower cost, moderate SO_2 emission reduction technologies will be important. In addition, changing technologies for generating electricity and producing combustion fuels present different situations for control and new opportunities for more cost effective control. As one looks at the emerging opportunites for long-range research in the control of sulphur oxides, one notes that there is a need to:

(1) Develop sulphur oxide emission control for stoker boilers, including the use of coal pellets;
(2) Demonstrate dry FGD on units burning high sulphur coals over a long operating period;
(3) Demonstrate dry FGD for coal–MSW mixtures to remove SO_2 and HCl simultaneously;
(4) Assess process changes to reduce sulphur oxides from secondary lead smelters;
(5) Assess sulphur oxide control strategies to reduce acid deposition problems;
(6) Perform operations research on combinations of sulphur oxide control technology for improved cost effectiveness;
(7) Seek innovations on existing technologies to greatly enhance effectiveness or greatly reduce cost;
(8) Develop/demonstrate lower cost retrofitable technologies, such as LIMB or other dry injection technologies applicable to existing coal-fired boilers;
(9) Develop cheap, high-reactivity alkali materials for dry injection post-combustion systems;
(10) Explore the use of additives to enhance spray-drier FGD technology;
(11) Develop/evaluate new innovative multi-pollutant control technologies which may yield much higher effectiveness or greatly reduced costs;
(12) Evaluate new/future energy technology and related sulphur controls to encourage opportunities for more effective control in areas such as gasification combined cycles for power generations, and fluidized bed combustion SO_x/

NO_x control innovations;
(13) Evaluate innovative chemical coal cleaning concepts to allow the sulphur to be removed prior to combustion.

3.2 Control of nitrogen oxide emissions

The increased use of fossil fuels, especially coal, in the future, will lead to increased emissions of nitrogen oxides from stationary sources. While few areas are now non-attainment for ambient air quality NO_2 standards, growth in NO_x emissions could increase the problem and defeat the EPA strategy to prevent future problems via increased control on new sources. Although all the answers are not known, NO_x appears to contribute substantially to acid deposition. Currently, demonstrating controls for NO_x are far behind other criteria pollutants in terms of removal efficiency. There is a need to develop efficient means of controlling oxides, in nitrogen (combustion modification and flue gas removal) from combustion sources (boilers, combustion engines, turbines, process heaters) and all other sources of NO_x emissions not now regulated (e.g. glass furnaces and cement kilns). Other areas of future and present concern relating to NO_x are increasing evidence that compounds such as nitro-aeromatics (PAN) can potentially contribute to atmospheric mutagenicity/carcinogenicity. Other nitrogen compounds emitted from combustion sources, like N_2O, have not been well quantified and should be characterized more closely until it is certain that they do not contribute significantly to potential problems such as ozone depletion. Also, NO_x controls need to be evaluated environmentally as they are developed to ensure that new, problem emissions are not being generated. Higher nitrogen-content fuels and high nitrogen-content hazardous waste are also projected to be used increasingly for fuels in the future and means for their effective, efficient combustion with low NO_x emissions are needed.

3.2.1 Current control methods

Reduction of oxides of nitrogen emissions in new or existing units may, in principle, be achieved through combustion modifications and/or flue gas treatment (FGT) technology. FGT systems with 80% NO_x reduction efficiencies have not yet been commercially demonstrated in the United States and are not widely regarded as a near-term option for NO_x control. Combustion modification techniques, however, may be capable of reducing NO_x emissions up to about 60% and are the principal methods being considered for NO_x control. These abatement approaches have been used on new fossil fuel boilers, and to a more limited extent on existing units. This abatement technology is basically used for controlling emission from oil or gas-fired units. There is very little experience of applying this technology on coal-fired units.

The abatement methodologies now used are based on lowering the oxygen content in the primary flame zone and on lowering peak flame temperature. The basic control methods using these two techniques can be classified in four distinct areas: (1) low excess air (LEA); (2) flue gas recirculation (FGR); (3) low NO_x burners (LNB); and (4) off stoichimetric combustion (OSC), often referred to as staged combustion. Stage combustion utilizes biased firing, burners out of service (BOOS), and over fire air (OFA).

LEA is an attempt to combust the fuel with a reduced excess of oxygen content in the incoming combustion air. Limited tests show about 100 p.p.m. NO_x reduction per 1% O_2 for wall-fired units and 10% reduction for tangential fire units. This abatement method is not recommended for cyclone units owing to corrosion problems. For coal-fired units, problems with LEA include smoking, slagging and excess char carryover (incomplete combustion) resulting in thermal efficiency losses. These deficiencies must be balanced against efficiency gains due to decreased stack heat loss.

In FGR control methods, about 15% of the flue gas is recycled to the secondary air input. Reduced NO_x emissions result from lower oxygen and lower peak flame temperatures. Very little data is available on this option, but information available indicates that a maximum of 20% NO_x reduction is possible.

Biased firing is a method which decreases the flow or increases the air to the upper burners in a combustion zone. This is perhaps the least effective control method with NO_x potential ranging only from about 5–7%. Another control method restricts the fuel load to certain burners in a combustion zone while the air flow is allowed to continue in order to maintain the proper heat content of the combustion chamber. The fuel flow to the other burners must be increased by a proportional amount. This is a very difficult option to implement in pulverized coal-burning facilities because the boiler may be derated in the process. By installing overfire air ports over the burner region and directing about 20% of the total combustion air through these ports NO_x reductions in the range 30–40% for tangential units and 30–50% for wall-fired units are possible. Finally, the most promising option for NO_x emission control are the low NO_x burners. These are potentially applicable for both new and retrofit units. When this option is feasible, it offers the highest degree of NO_x control potential with 40–60% reductions for coal-fired boilers.

The cost of control for oxides of nitrogen control methods is variable. Figure 7, however, does give some indication of the total capital cost of NO_x control as a function of plant capacity and method of control. It should be noted that the total capital costs of NO_x control are quite small compared with other pollution control costs, typically in the order of 5% of the cost of FGD systems. Operating costs are often negligible.

3.2.2 Emerging control methods

The research agenda for long-range research for the control of nitrogen oxides is long. The list includes projects to:

(1) Define all the physical and chemical phenomena contributing to NO_x formation and destruction in combustion processes;
(2) Develop improved models to predict the performance and cost of combustion modification of NO_x and air toxics;
(3) Develop inexpensive reliable continuous NO_x monitoring equipment;
(4) Develop more advanced effective NO_x controls which are applicable and retrofitable to several key sources, e.g. reburning applied to oil and coal combustion and pre-combustors applied to oil and high nitrogen waste;
(5) Develop specific combustion modification applications for several key sources (small industrial boilers, stationary engines, industrial process furnaces, package boilers);
(6) Investigate interactions in NO_x controls for energy efficient equipment such as

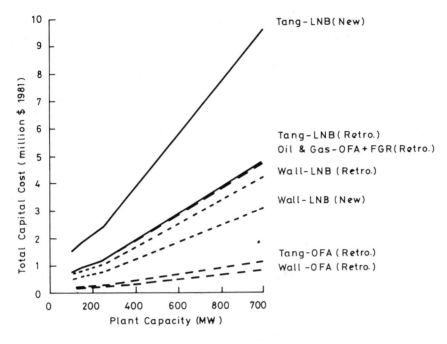

Fig. 7 — Total capital costs of NO_x control (interest during construction is not included).

gas turbine/duct burners and gas turbine/boiler combinations;
(7) Develop, demonstrate and evaluate effective selected catalyst control devices for stationary engines, including gas-fired internal combustion engines, diesel engines, and gas turbines;
(8) Demonstrate and assess the effectiveness of combined control approaches for maximum NO_x reduction to selected combustion sources.

3.3 Control of volatile organic compounds

Another class of gaseous compounds under active consideration for controls is volatile organic compounds (VOC). VOCs are of particular interest because they are believed to be the precursors to photochemical oxidants. In 1975 nineteen million tons of VOCs were emitted from stationary sources which represented 61% of the national total. The categories of sources of VOCs include solvent evaporation (40% of national VOC emissions) and organic chemicals (1% of national VOC emissions). Of the total national VOC emissions, 80% are from stationary sources. The primary control technologies include the use of flares, both catalytic and thermal incineration, and carbon absorption. Recently a series of tests have been conducted on a pilot scale catalytic oxidizer controlling VOC emissions from a flexographic printing process. The study evaluated the effect of flow rate and temperature on the VOC destruction efficiency. The study found that over a period of five months of operation that if the space velocity was kept at 50 000 FPM or below, and the temperature above 315°C, efficiencies of 90% or better were achieved during the entire operation.

The organic compound being controlled was n-propyl acetate. Another test examined a full-scale catalytic oxidizer installed on a formaldehyde production unit. Results of these tests indicated that the overall destruction efficiencies were 80% or better. Yet another project examined field evaluations of industrial catalytic oxidizers at six sites (eight incinerators). Results of these tests showed that equipment met design specifications except when catalyst deactivation had occurred. EPA conducted flare efficiency tests on an 8-inch steam assisted flare at the John Zinc Company test facility. The test programme evaluated the effect of flow rate, field heat content and steam injection rate on the overall combustion efficiency. The combustion efficiency during the test generally exceeded 98% except under conditions of excessive steam and low heat content/high flow. Finally the EPA designed, constructed, and operated a flare test facility to evaluate flare efficiency and emissions for 3-, 6- and 12-inch flare heads. The test facility included provision for steam injection, fuel mixing, tracer injection and multi-point sampling. The initial programme has been completed and included 75 tests covering a range of waste gas flows, exit velocities, heat contents and steam injection rates. The tests were also conducted on three flare heads supplied by flare manufacturers. Although the final report of these tests is not yet available, typical results for the 3-inch flares are given in Table 5.

3.4 Simultaneous removal of SO_x and NO_x

The cornerstone of the EPA's programme to simultaneously reduce SO_x and NO_x emissions in one unit is the limestone injection into a multi-state burner (LIMB) project, an effort to develop an effective and inexpensive emission control technology for coal-fired boilers. The LIMB technology represents a low-cost alternative to currently available SO_x control approaches. LIMB is especially attractive if coal combustion must be controlled to minimize emissions of acid rain precursors because LIMB is: (1) easily retrofitted to large and small coal-fired boilers; (2) potentially the lowest cost alternative; and (3) capable of controlling both SO_x and NO_x emissions. Major research programmes to develop direct limestone injection are being sponsored by EPA and EPRI.

The technical goals of the EPA research programme are:

(1) For retrofit applications to achieve 50–60% reductions of both SO_x and NO_x;
(2) For new systems to achieve 70–80% NO_x and 70–90% SO_x reductions; and
(3) For both retrofit and new systems to achieve the above goals at costs at least $100/kW less than the major technology alternative, flue gas desulphurization.

The EPA development programme to achieve these goals consists of:

(1) Developing a basic understanding of the mechanisms and kinetics of the process;
(2) A systematic small bench and pilot scale development programme;
(3) Large-scale pilot testing;
(4) A detailed process and systems analysis for the process;
(5) Field application to representative boilers.

The EPRI programme emphasizes combining limestone injection with less costly and retrofittable combustion modification techniques which can be applied to

Table 5 — Global combustion efficiency of 3-inch EER flare head at high velocities [1]

Purpose of test	Test no.	Flame retention ring	Actual exit velocity (ft/s)	Nominal exit velocity (ft/s)	Low heating value (Btu/ft³)	Fuel in nitrogen (%)	Steam ratio (lb. steam/ lb. fuel)	Wind speed (mph)	Flame length (ft)	Lift off (in)	Observations Colour	Smoke	Sample method (R. rate; H. hood)	Probe position (ft)	Global combustion efficiency (%)
High vel. no. flame net ring	77	No	39.6	39.6	1316	56.0	0.0	0–3	24	5–9	Yellow	Yes	R	25	95.11
	78	No	80.0	80.0	1325	56.4	0.0	—	28	12–13	Yellow/orange	Yes	R	33	98.40
	79	No	118.5	118.5	1307	55.6	0.058	1–5	32	18	Yellow/orange	Yes	R	35	99.66
High vel. 56% C₃H₈	80	Div. 40.0%	420.5	168.2	1255	53.4	0.015	0–1	33	24	Clear/orange	No	R	34	99.20
	81		67.0	26.8	837	35.6	0.149	0–1	17.5	8–10	Orange	No	R	20	97.27
	82		248.3	99.3	1107	48.1	0.031	0–1	31	24	Orange	No	R	35	99.33
High vel. 100% C₃H₈	83	Div. 40.0%	144.5	57.8	2350	100.0	0.131	0–1	29.5	12	Orange	No	R	33	99.87
	93	Conv. 46.3%	84.7	39.2	2348	99.9	0.122	0–5	26.5	10	Orange	No	R	31	99.74
	94		171.4	79.3	2350	100.0	0.061	0–3	34.5	12	Yellow/orange	Little	R	37	99.83
High vel. stable flame limit	95	Conv. 46.3%	114.8	53.1	792	33.7	0.087	0–2.5	23	12–18	Yellow	No	R	23	99.87
	96		274.1	126.8	1043	44.4	0.028	0–1.5	33	18–28	Yellow/orange	No	R	37	99.85
High vel. 50% C₃H₈	97	Conv. 46.3%	85.8	39.7	1156	49.2	0.142	1–5	20	8	Orange	No	R	24	99.72
	98		172.3	79.7	1133	48.2	0.071	0–2	29	7	Yellow/orange	No	R	34	99.87
High vel. stable flame limit	99	Conv. 46.3%	428.2	198.1	1027	43.7	0.018	0–1	30.5	24–36	Orange	No	R	35	99.88
High vel. 50% C₃H₈	100	Conv. 46.3%	343.1	158.7	1149	48.9	0.020	0–4	34	24–30	Orange	No	R	37	99.84
High vel. 77% C₃H₈	101	Conv. 46.3%	85.8	39.7	1807	76.9	0.091	1–4	25	6	Orange	No	R	29	99.73
	102		170.1	78.7	1805	76.8	0.80	1–5	30.5	12	Yellow/orange	Little	R	34	99.74
	103		260.1	120.3	1795	76.4	0.060	0–1.5	37	8	Yellow	No	R	37	99.88
High vel. stable flame limit	104	Conv. 46.3%	375.5	173.7	921	39.2	0.023	0–1.5	29.5	36–48	Orange	No	R	30	99.81
High. vel. 100% C₃H₈	105	Conv. 46.3%	243.4	112.6	2350	100.0	0.049	0–5	40	12	Orange	Little	R	42	99.77

existing burners to obtain similar levels of SO_2 control but half the level of NO_x control. After review of past and current pilot plant results, plus engineering evaluation for commercial units, results from both programmes show attractive costs for SO_2 removal in the 50–60% range.

The LIMB capital costs are made up of four components: (1) reagent material handling and preparation; (2) boiler modification; (3) particulate removal; and (4) waste disposal. Except for boiler modification, the capital costs for the remaining components are identical to the conventional lime/limestone FGDs.

Boiler modification costs can be broken down into two components. The first is the limestone injection mechanism, estimated by TVA to be $6 kW (1983$) for a 200 MW boiler. The other component includes boiler modifications necessary to obtain low NO_x combustion. This can vary considerably depending on the extent of modifications required. In a recent study where only stage combustion modifications were required on a tangential boiler, the cost was estimated at $1/kW (1982$) for 500 MW units. In another study where new burners were estimated for retrofit on a 200-MW wall-fired boiler, the cost was $16/kW (1983$) given by Babcock and Wilcox as an order of magnitude estimate. An average estimate for boiler modification costs for LIMB are of the order $10/kW (1980$). The costs of modification are partitioned as follows: limestone injection costing approximately $6/kW plus $4/kW for boiler modifications.

REFERENCES

[1] Private communication, EPA.
[2] PEDCo Environmental, Utility FGD Survey, EPRI Contract No. RP982-32, July–September 1983.

2.3

Economic issues in the control of air pollution†

I. M. Torrens
OECD, Resources and Energy Division, 2 rue Andre Pascal, 75775 Paris, France

1. MAJOR AIR POLLUTION ISSUES

Three decades ago, as a consequence of air pollution crises in urban areas involving considerable numbers of premature deaths, it became evident that measures needed to be taken urgently to clean the air in cities to a level where humans could breathe safely. There is still some debate as to exactly what this level is, especially since some people have less robust respiratory systems than others. However, since the 1950s we have witnessed a gradual tightening of pollution control standards in our cities, and a gradual improvement in the quality of their air.

What goes up must normally come down, however, and, as a result of the use of high stacks to disperse air pollutants, part of the improvement in urban air quality has come at the expense of damage elsewhere. A corollary of this is that a reduction of health risks to man has been bought at the expense of an increase of health risks to our natural environment. Parts of our man-made environment have also been found to be sensitive, over a longer period, to levels of air pollution which are not judged to be harmful to human populations. Some of the damage to property can be dealt with by increased maintenance (like painting bridges and other metal-work), but damage to historical monuments or stained-glass windows cannot be so easily remedied and, once destroyed, ancient cultural artefacts are not possible to replace.

There are many different types of air pollution issues, including toxic substances in the atmosphere (mainly carcinogens and trace metals), or respirable fine particulates. This paper, however, limits itself to the major gaseous air pollutants — mainly sulphur oxides (SO_x), nitrogen oxides (NO_x) and hydrocarbons (HC), the last being more correctly termed volatile organic compounds (VOC). Another major air

---

† The opinions expressed in this chapter are those of the author and do not necessarily represent the views of the OECD or of the governments of its member countries.

pollutant, particulate matter (PM), was a source of greater concern but is now subject to quite stringent controls in most OECD countries.

Both stationary combustion sources (power plants, industrial installations) and vehicles emit these pollutants. SO_x is emitted almost entirely from stationary sources, whereas NO_x and HC emissions are split approximately equally among stationary and mobile sources in most industrialized countries. Perhaps the key air pollution issue today is acid rain, or more correctly acid deposition, including wet and dry forms [1]. Following emission of the gaseous pollutants, sulphur and nitrogen compounds formed in the atmosphere through chemical transformation can travel long distances — hundreds or even thousands of kilometres — and return to the earth as rain, snow or dry deposition. This can affect lake ecosystems, forests, crops, materials and human health (the last mainly via drinking water).

Initially, attention arising from concern about acid rain was focused almost entirely on sulphur oxides and measures have been taken in a number of OECD countries to reduce SO_x emissions. More recently, nitrogen oxides have attracted increasing attention. They are estimated to be responsible for about 30% of the acidity of deposition. In addition, while fuel substitution and other pollution control measures, as well as industrial restructuring, have resulted in stabilization or even a decrease in SO_x emission in a number of major industrial countries over the past decade, emissions of NO_x are continuing to rise.

The secondary pollutants arising from NO_x emissions and atmospheric interactions include not only nitric acid but also (through interaction of NO_x and HC in the presence of sunlight) photochemical oxidants, mainly ozone. The latter is increasingly thought to be implicated in air pollution damage to plants and trees and even to materials.

Controlling the effects of air pollution has traditionally meant dispersing it over a wider area, for instance through the use of tall stacks. However, since the problems of environmental damage from acidification often arise far from the principal sources of these pollutants, it is clear that dispersion is not an adequate control method. Apart from some remedial measures which can be taken (liming of lakes, for example), this leaves the reduction of pollutant emissions as the principal means of control.

2. EMMISION STANDARDS FOR MAJOR AIR POLLUTANTS IN OECD COUNTRIES

All OECD countries exercise some form of environmental control over airborne emissions resulting from electricity generation. In some cases, this control is also applied to industrial boilers. However, there is a wide variation among countries with respect to both the level and the type of regulation employed. Indirect control of emissions can be exercised, for example, by limitations on the sulphur and ash content of fuel imports and the application of fuel quality standards with the country.

The specific pollutants subject to quantitative emission limits reflect, *inter alia*, the availability of proven control technology. Particulate emission control has been widely practised for many years and is proven in terms of reliability and efficiency.

Control equipment to reduce SO_x and NO_x is now considered to be commercially available and proven, but is, comparatively speaking, more recent and less widespread in application. The number of countries enforcing a quantitative standard for these pollutants reflects this. Table 1 indicates the quantitative emission standards currently in force in a number of OECD countries.

Where countries do not appear in Table 1 (e.g. France or Italy for sulphur oxide control) this does not necessarily mean that no restrictions apply. There may be emission limits applying on a local or regional basis, or sulphur emissions may be controlled in some other way, e.g. through limits on the sulphur content of fuels.

3. TECHNOLOGIES FOR EMISSION CONTROL

Pollution control can involve more than just preventing gaseous emissions from leaving the smoke stack or car exhaust. The choice of fuel, the treatment it undergoes prior to combustion, the combustion process itself, and the cleaning of combustion gases before emission, can all contribute to a greater or lesser extent (see Table 2).

Reliable technologies and methods exist for achieving a high degree of control of the major air pollutants from most sources: more than 99% of particulates and hydrocarbons, up to 95% of sulphur oxides, and 90% of nitrogen oxides can be removed from flue gases in large installations. Some of these technologies have been developed quite recently and were either not available (flue gas denitrification) or were considered not sufficiently reliable or too expensive (flue gas desulphurization) to be installed in large combustion plants in most OECD countries in the past.

The relative costs of pollution control in smaller plants used by industry are more onerous than for large power plants, as they derive less benefit from economies of scale. To date, these plants have relied for pollution control less on combustion and post-combustion clean-up than on choice of fuel and fuel treatment. Here, there are indications that new technologies, recently initiated and now under development, may hold good promise for the future. Atmospheric fluidized bed combustion, now being offered commercially by a number of manufacturers, gives smaller boilers the capability of up to 90% sulphur removal in the combustion process. Another new technology, limestone injection in multi-stage burners, again aimed at reduction of sulphur and nitrogen oxides leaving the combustion chamber, is presently at the pilot plant stage. Both of these technologies are being scaled up in test facilities in the United States and the Federal Republic of Germany, with the objective of assessing the possibility of applying them to electricity generation plants.

As far as mobile sources are concerned, pollution control is mainly aimed at nitrogen oxides and hydrocarbon emissions. The catalytic converter, now used in the United States and Japan, reduces emissions of both by up to 90% (and requires lead-free fuel to avoid destroying the catalyst). Other technologies to reduce one or other of these pollutants (e.g. the 'lean-burn' engine or stratified charge engine) are also aimed at increasing fuel efficiency — itself a means of reducing pollutant emissions. Judging from past experience, the requirement to meet lower pollutant emission standards can be a powerful stimulus to develop more efficient and effective technologies for both combustion and exhaust-gas clean-up.

Table 1 — Comparison of national emission standards for electricity generating plants (Taken from *Emission Standards for Major Air Pollutants from Energy Facilities in OECD Member Countries*, OECD, Paris, 1984)[a]

Country	Fuel	Emission limit[b]		Comments
		(mg/Nm3)	(ng/J = g/GJ)	
A. Particulate control				
Australia	Solid	250	105	National guidelines
Belgium	Solid	350	147	Regulations
Canada	All	116	43	National guidelines
Denmark	Liquid	97	36	National guidelines
	Solid	150	63	
Germany	Liquid	50	18	Regulations
	Solid	50	21	
	Gas	5	2	
Greece	All	150	56	Regulations
Japan	Liquid	50	18	Regulations
	Solid	100	42	
	Gas	50	15	
The Netherlands	Solid	48	20	National guidelines
New Zealand	Solid	125	52	Regulations
Sweden	Solid	36	15	Regulations
United Kingdom	Solid	115	48	
United States	Solid	31	13	Regulations
B. Sulphur oxide control				
Belgium	Liquid	5000	1850	Regulation (normal conditions)
		2000	740	Regulations (alarm conditions)
Canada	All	700	258	National guidelines
Germany	Liquid	400	168	Regulations
	Solid	400	148	
	Gas	35	10	
Japan[c]	Liquid	549	203	Regulations
	Solid	549	230	
	Gas	549	165	
The Netherlands	Solid	548	230	National guidelines
Sweden	Solid	240	100	Proposal
	Liquid	270	100	
United States	Liquid	920	340[d]	Regulations
	Solid	1238	520[e]	
	Gas	920	340[d]	

Table 1 — *Continued*

Country	Fuel	Emission limit[b]		Comments
		(mg/Nm3)	(ng/J = g/GJ)	
C. Nitrogen oxide control				
Australia	Gas	*350*	105	National guidelines
Canada	Liquid	350	*129*	National guidelines
	Solid	614	*258*	
	Gas	287	*86*	
Germany	Liquid	*450 (150)*	167 (56)	
	Solid	*800 (200)*	333 (83)	Regulations
	Gas	*350 (100)*	106 (30)	(Proposals)[g]
Japan	Liquid	*267*	99	Regulations
	Solid	*616/411[f]*	259/173[f]	
	Gas	*123*	37	
The Netherlands	Solid	643	*270*	National guidelines
Sweden	Solid	667	*280*	Regulations
United States	Liquid	570	*210*	Regulations
	Solid	619	*260*	
	Gas	287	*86*	

[a] Considerable variation occurs among countries in both definition and conditions of application of emission standards. Comparisons should be made with caution, and readers should refer to specific country tables of the compendium for further information.
[b] Italics indicate units in which limits are normally expressed in specific countries. Conversion between mg/Nm^3 and ng/GJ is carried out using the following conversion factors: 420 m^3/GJ (solid); 370 m^3/GJ (liquid); 300 m^3/GJ (gaseous fuel); for flue gas at 1 Bar pressure and 15°C (12% CO_2 content of flue gas for solid and liquid-fired boilers).
[c] Assume effective stack height 260 m K-value 3.0, and volume of flue gas 900 000 Nm^3/h (see country notes for Japan). These values would typically apply to a power plant in an urban area.
[d] Ten per cent of potential combustion concentration (90% reduction), or no reduction when emissions are less then 86 ng/J.
[e] Ten per cent of potential combustion concentration (90% reduction) or 30% of the potential combustion concentration (70% reduction) when emissions are less than 260 ng/J (solid fuels).
[f] Facilities installed before/after 31 March 1987.
[g] Maximum values according to ordinance (or envisaged figures to be applied in the licensing procedure).

4. EMISSION CONTROL COSTS

4.1 Costs of control technologies

With a wide range of possible control technologies and methods applicable to an equally wide range of emitting installations, there is clearly a high degree of uncertainty in any estimate of the control cost. One fundamental difficulty in assessing environmental control costs is the question of what may specifically be termed 'environmental control'. For instance, in a power plant, in the case of add-on emission control equipment, like a flue-gas desulphurization system or an electrostatic precipitator, the attribution of costs to environmental control is relatively simple. But in other cases, such as wastewater, thermal effluent and noise control, it is sometimes unclear in reported cost data whether these should be included in environmental costs or are part of the general plant costs as they would be done even in the absence of regulation. The same is true of buildings, conveyor belts and other

Table 2 — Air pollutant emission reduction technologies

Pollutant	Control technology	Effectiveness
Stationary sources		
SO_x	Fuel selection	Depends on S content
	Coal cleaning	Depends on fuel quality: up to about 30% S removal for coal.
	Distillate oil desulphurization	Can reduce S content to 0.15% by weight
	Fuel oil desulphurization	Usually not competitive with FGD
	Limestone injection in combustion chamber	Up to about 60% removal
	AFBC	Up to 90% S removal
	Flue gas desulphurization (FGD)	Up to 95% SO_2 removal
NO_x	Combustion modifications	Up to 60% NO_x reduction
	Flue gas denitrification (selective catalytic reduction)	80% NO_x removal
HC	Condensers	85–99% HC reduction
	Carbon adsorption	35–99% reduction depending on process
	Incineration	90–99% HC reduction
Mobile sources		
NO_x	Exhaust gas recirculation (EGR)	50% reduction
	Lean combustion with EGR	Up to 85% reduction
	Three-way catalyst	50–80% reduction
HC	Lean combustion with EGR	25–50% reduction
	Three-way catalyst	60–90% reduction

structures which are designed to minimize noise, dust and other nuisances at the power plant. Care needs to be exercised, therefore, in interpreting and, in particular, comparing environmental control costs for different plants. Important differences in scope and definition should, as far as possible, be made explicit.

Other problems in comparing cost estimates for specific pollution abatement technologies arise because of differences in methodologies and specific assumptions employed by different authors and organizations. For example, air pollution abatement costs depend on a number of key physical parameters (related primarily to fuel characteristics, power plant design, and applicable regulatory standards), as well as on various economic parameters that strongly affect capital and operating cost calculations. Lack of a standardized methodology and nomenclature for identifying and reporting all elements affecting cost calculations inevitably hinders the ability to assure that any comparisons made are systematic, and reflect a consistent set of premisses.

Recognizing these difficulties, it is still possible to highlight the relative importance of control costs for different air pollutants. This chapter claims to be indicative rather than comprehensive, selecting as a specific example the case of a new coal-fired power plant in an OECD country and giving some cost ranges drawn from recent experience and estimates (Table 3).

Table 3 — Indicative air pollution control costs in a new baseload coal-fired power plant (1984 US$)

Pollution and control method	Capital cost ($/kW)	Annual total cost (capital and operating)		Operation and maintenance (% of annual cost)
		(mills/kW)	(% of cost of electricity generated)	
Particulate control				
ESP	15–40	1–3	1–3	25–55
Fabric filter	25–50	2–4	2–4	
Sulphur oxide control				
FGD (entire flue gas stream)	70–185	5–12[a]	9–16[a]	35–70
Nitrogen oxide control				
Combustion modifications	5–15	1–2	1–2	
Flue gas denitrification[b]	35–85	2–6	3–5	50–80

[a] The range given here for FGD is very broad. The capital costs reported in OECD countries for FGD at 90% efficiency of sulphur removal applied to the entire flue gas stream mostly fall in the range of $140–185/kW for the USA, $90–150/kW of installed generating capacity for Japan, and $70–130 for Europe.
[b] Experience in flue gas denitrification is still limited, being applied in full-scale coal-fired installations only in Japan.

Recent work in OECD on coal pollution abatement [2,3] has highlighted the difficulties involved in estimating the costs of pollution control, and particularly in comparing estimates across countries. There are many uncertainties. The problems are perhaps best illustrated by the case of flue gas desulphurization, the most costly of the control technologies and the one with the widest range of estimated costs (see Table 3). A careful analysis of the many different factors, both hardware and financial, which enter cost estimates in several major OECD countries, has revealed important real differences in the capital cost of FGD in the United States compared with Europe and Japan. While the strength of the US dollar and the higher real interest rates in comparison with the currencies and interest rates of other countries are significant factors, a major part of the cost variation appears to arise from very different approaches to the configuration of the FGD systems in the United States compared with elsewhere. Early difficulties with FGD reliability combined with regulatory requirements led the United States to develop the modular approach, with typically, for a 600 MW power generating unit, four 150 MW FGD units plus one spare unit (i.e. 20% redundancy). This configuration is still the practice in the United States, though many of the reliability problems have been largely solved through experience and improved maintenance and control.

In Europe and Japan, on the other hand, the practice adopted has been to build single FGD units capable of handling the entire flue gas stream, with no, or only a token amount, of redundancy. This leads to very significant economies of scale in construction (one unit instead of five), as well as reductions in construction time (hence interest during construction). Other differences include engineering contingency charges (high in the United States and low elsewhere). What is emerging is a rather different range of FGD costs for the United States situation and for other major OECD countries (Table 3).

To go into this degree of detail on the specific item of international FGD cost comparisons may seem unnecessary but it must be remembered that FGD is a highly significant item in the current international debate on air pollution, mainly because of its reputedly high cost. Consequently, any information or insight which indicates the possibility of a sizeable reduction in its cost could alter the terms of reference of that part of the debate.

The problems associated with estimating the costs of pollution control applied to vehicles are, if anything, more complex. If emission reductions are achieved through specific add-on equipment, the cost situation is clearer: the three-way catalytic converter to reduce emissions of NO_x, HC and CO costs approximately $300–500 per car, when other necessary modifications to the propulsion system are taken into account.

A great deal of research and development by the motor industry has been devoted over the past decade towards the simultaneous reduction of the amount of pollutant emitted and the vehicle fuel consumption. It is difficult to assess how much additional cost to attribute to the gradual improvement in vehicle energy/environment performance, and particularly how much of this cost should be allocated to pollution reduction. The 'lean-burn' engine is now being developed in a number of countries as a promising way to meet stricter emission standards, especially for smaller cars. It reduces emissions of NO_x but not of unburnt HC. However, in association with an oxidation (simpler than a three-way) catalyst, significant reductions in both NO_x and HC could be achieved. How much of the cost of a lean-burn engine is an environmental control cost is an almost impossible question to answer.

4.2 Cost of national or international emission reductions

If the estimation of pollution control costs is complex for a single plant or vehicle, a much higher degree of uncertainty applies to estimates of the cost of acid rain control strategies at the national or international scale. Among the sources of uncertainty, even given the constancy of environmental regulations, are the difficulty in assessing the future economic growth or electricity demand; the rate at which new generating capacity comes on stream and old capacity is phased out; and the difficulty in predicting when technological improvements will be available for commercialization and at what cost.

Notwithstanding these difficulties, overall cost estimates have been made for control strategies for SO_x in Europe and the United States. These include the installation of pollution control equipment on both new and existing power plants and industrial installations.

Table 4 gives an idea of the order of magnitude of such costs. These are, of course, large numbers. But we should try to see them in relation to what they would mean for the electricity consumer, for example. In the German case, the SO_2 reduction given in Table 4 is estimated to add up to one pfennig per kW/h of the electricity price to consumers (or approximately 6%). The United Kingdom Central Electricity Generating Board has estimated that retrofitting FGD to existing coal-fired power plants to achieve a 50% reduction might add, over a period of more than a decade, about 5–6% in total to the average cost of electricity generated in the UK and considerably less to consumers' electricity bills. These are not negligible increases, but they are far from catastrophic.

This brief overview of the costs side has not addressed the costs of control for mobile sources of air pollution. In fact, as mentioned earlier, a very different situation prevails in the different regions of OECD, and it would be difficult to come up with any comparable numbers for overall control costs.

According to the US Department of Commerce [10], all controls of NO_x, CO, HC, and PM from mobile sources cost US $16.5 billion annually in 1981 (1981 $). These costs are the total for the approximately 136 million light- and heavy-duty vehicles in the United States in that year, or an average of £121 per vehicle per year. (That figure, which seems high, covers all costs of the mobile source air pollution programme, including programme administration and enforcement, fuel maintenance, etc. added by the controls but does not include research and development costs.) This has achieved a reduction of over 90% of emissions from auto gasoline engines and has reduced diesel emissions as well.

5. ECONOMIC ISSUES RELATING TO AIR POLLUTION CONTROL

The previous sections have set the stage for a discussion of the economic issues. The crux of the present policy debate is: in the light of the potential costs of control measured against the potential benefits in the form of avoided damage to the environment, is further action to reduce air pollution justifiable and necessary? A great deal of debate is focused on the question of whether to incur the substantial costs associated with a reduction in emissions of major air pollutants even though the cost/benefit picture is not clear.

This introduces a number of subsidiary issues.

5.1 The appropriate level of control for different types of source

As mentioned earlier, pollution controls are often more expensive for smaller combustion installations than for larger plants. If this is the case, and if alterations such as better and less polluting combustion technologies are not available, there may be a case for applying less stringent emission standards to industrial boilers than to power plants. Whatever the standards are, however, they should reflect what is achievable by the best technologies that are economically feasible for such plants.

For mobile sources the picture is more complicated. Catalytic converters on

Table 4 — Costs of strategies to control air pollutants

Country or group	Air pollutant	Emission source (stationary)	Reduction	By year	Annual cost	Source
European Community	SO_2	All	10–13 M tonnes (53–77%)	2000	US$4.6–6.7 bn[a,b]	Ref. 4
	NO_x	All	50%	2000	US$0.4 bn[a]	Ref. 4
Federal Republic of Germany	SO_2	All	1.6 M tonnes	1993	DM3.3 bn	Ref. 5
United Kingdom	SO_2	Power plants	1.5 M tons (75%)	1995+	£175 m[c] (average)	Ref. 6, 7
United States	SO_2	Power plants	10 M tons (75%)	1995+	US$4.2–5.3[a]	Ref. 8
United States	SO_2	Power plants	10 M tons (75%)	1995+	US$5.2–9.5[a]	Ref. 9

[a] 1982 prices
[b] Includes retrofitting FGD on 70% of large coal and lignite boilers over 25 MW.
[c] 1983 prices. Includes cost of replacement and output capacity lost through retrofitting of FGD. This accounts for about 25% of total capital costs of £1990 million.

smaller cars add a much larger percentage to the purchase price then they do for larger cars. But, in addition, smaller cars, uncontrolled, emit lower amounts of pollutants than do larger ones. A lot depends therefore on the emission limits set and on the most economical way that cars of different sizes can meet them. In fact, the choice of emission limit levels is crucial to the choice of technologies, and thus to the overall cost. In the 1970s in the United States and more particularly in Japan, governments set emission limit targets for future years which turned out to be 'technology-forcing' and which were met by development of the catalytic converter and improvements in fuel efficiency.

5.2 How to treat new emission sources and existing sources?

A perennial problem in industrialized countries is how to deal with polluting installations which were built when control regulations were less strict or non-existent. Large installations in particular may have several decades of useful life still ahead, and yet may emit levels of air pollutants far in excess of those required of a similar new plant; for example, some older coal-fired power plants in the United States emit 10–15 times the amounts of SO_2 which an equivalent, new plant would be allowed to emit under the federal new source performance standards.

The philosophy adopted in OECD countries until recently was that it is not appropriate to apply pollution control regulations retroactively, requiring older plants to retrofit new equipment. However, new regulations in the Federal Republic of Germany and the Netherlands have introduced retrofit requirements under certain conditions. A lively part of the current debate on air pollution concerns whether or not this approach should be adopted by other countries. Certainly, in a number of countries it would be difficult to achieve large reductions in emissions, particularly of SO_x, without a degree of retrofitting.

5.3 Who pays for increased controls?

There is an easy and a difficult answer to this question. The easy one is to apply the OECD Polluter Pays Principle, which states that the costs of preventing or controlling pollution should be borne by the polluter. For example, installing FGD in a power plant adds to the costs of the generation of electricity. Normally the additional costs will be reflected in the electricity price to consumers.

However, the Polluter Pays Principle was not designed to cover pollution at long range, which can occur following emissions of air pollutants, especially from tall stacks. In the above example, the electric utility may be far in distance from areas where the greatest environmental effects are perceived, and if the cost of reducing pollutant emissions is passed through to electricity prices, the consumers who pay these increased prices are in many cases not the same as those who presently bear the costs of environmental damage (often they are not even citizens of the same country).

A further complication of increased pollution control requirements is that they can place greater burdens on one region or country than on another. In some cases this may be because that region or country tolerated more polluting practices than its neighbours in the past. But it can also be related to structural factors (more manufacturing industry) or patterns of fuel consumption. For example, according to some bills currently before the US Congress, a number of utilities in the mid-West and South East, which depend on medium and high sulphur coal, would be required to incur greater pollution control costs (via retrofitting) than other utilities. This is estimated to have adverse effects, via electricity rate increases, on the region's manufacturing industry, much of which has already suffered greatly through loss of competitiveness in recent years. This is why some proposed bills in Congress include other suggestions on means of payment, such as a tax on all electricity generated in the United States.

In some countries, where electricity is a nationalized industry and priced nationally, the cost of proposed pollution reduction measures, when blended into the system, has a considerably smaller impact (usually only a few per cent) on the price of electricity to consumers.

5.4 How to achieve the most economically effective results?

An unwritten principle of pollution control is that it is usually more cost-effective for regulations or standards to specify the end but not the means, e.g. to set the emission limits but not to specify the technologies or methods used to meet these limits. Of course such a clear distinction is not very realistic in practice: emission limits must be fixed with an eye to the possible and economically feasible, which in turn demands some appreciation of what can be achieved by present technology and at what cost.

There are several different levels of specificity, however, ranging from a single plant or single vehicle emission up to a total national emission of a given pollutant. As we go up the scale, the degree of flexibility which is, in theory, possible in responding to a policy objective of pollution reduction becomes somewhat greater. An example of this is the so-called 'bubble concept' which has been applied on a modest scale in the United States. According to this, emission sources of a single company (or within a limited geographical area) are considered to be enclosed in a conceptual bubble. A target emission reduction, or a total emission limit based on some measure of production or fuel use is set, and it is left to the operator of the installations to decide how best to comply with it. A typical response might be to install a high degree of pollution control on one outlet and leave others unaltered. When there is more than one operator, the bubble can be accompanied by some form of 'emission trading' whereby one installation may find it more economic to continue former levels of emission and pay another operator part of the cost of installing pollution control equipment.

6. BENEFITS OF AIR POLLUTION CONTROL

No analysis would be complete without referring to the costs of not controlling these air pollutants, or of doing so to an inadequate degree. The issue of benefits of pollution control is a very complex one, because of the great diversity of the field environmental damage to be covered; the uncertainties in linking (particularly in a quantitative way) air pollution emissions, or even ambient concentrations, with damages produced; the difficulties (and value judgements involved) in expressing these damages in monetary terms; and the intangible nature of some of the effects (e.g. damage to historical monuments) which makes conversion to monetary terms impossible in such cases.

To cite the figures from some of the major benefits studies which have been carried out, including one by the OECD published in 1981 [11] and a more recent one carried out for the European Community by Environmental Resources Limited [4], with all the necessary qualifying cautionary statements would require a separate chapter. Suffice it to say that an increasing number of serious studies suggest that the costs and benefits of substantial reductions in pollutant emissions could be of a similar order of magnitude, and that a significant fraction of the benefits do not lend themselves to easy quantification.

A second economic factor on the benefits side, which is not often taken into account in the policy debate, is the stimulative impact which stricter environmental standards can have on economic activity and on technological development. Research and development and manufacturing pollution control equipment, contribute to our economic progress, and in particular to GDP. They can also add to levels of employment in sectors which are at present working well below capacity. Finally, they can result in lowering both environmental control costs and other costs through technological advances (e.g. development of more fuel-efficient automobiles).

7. CRITERIA FOR POLICY ACTION: LIVING WITH THE UNCERTAINTIES

It will be clear from the preceding sections that economic considerations are of prime importance to the current debate on the need for further policy action to reduce air pollution. Two basic policy stances are held by different groups of countries. The first, which has gathered a substantial number of new recruits over the past two years, holds that while the scientific cause/effect picture is admittedly not fully clear, the evidence linking emissions of acidifying air pollutants and important damage, particularly to forests, lakes and materials, is sufficiently strong to justify action now, even at a substantial cost. The opposing school of thought holds that the curbs to utilities and manufacturing industry would be unacceptably high, to the point where the regional or national economy would be adversely affected by a decision to reduce emissions substantially. This school of thought holds that such increased costs should be avoided until and unless the environmental case is proven by further research and assessment.

The policy dilemma illustrated in simplified terms in Fig. 1 poses an important question for policy makers. In view of the complexity of the processes involved, can we expect further research and assessment alone to inform us reliably of the extent to which a reduction in pollutant emissions will be effective in reducing environmental damage? Can we in fact establish the cost/benefit balance adequately without 'getting our feet wet' and taking some action to reduce air pollution by an amount which should in principle be translated into reduced damage? Policy makers in the environmental field often have to come to terms with uncertainty in their decision-making and to take into account many factors including those outside the environmental field, particularly economic factors.

Perhaps the area where there is most agreement among countries is the urgent need to disseminate reliable information on current and developing pollution control technologies and less polluting processes, as well as to work towards lowering the cost of pollution control technologies. If real progress can be made rapidly towards more cost-effective control, it would have an important effect on the economic parameters of the policy debate.

Part 3
Water treatment

3.1

Megatrends in water treatment technologies

F. Fiessinger, J. Mallevialle, A. Leprince and M. Wiesner
Laboratoire Central de la Lyonnaise des Eaux, 38 rue du President Wilson, 78230 Le Pecq, France.

1. INTRODUCTION

The objective of this chapter is to help reach a better understanding of the possible evolution of water treatment in the coming years. Predictions of the future are always controversial and this one will undoubtedly raise numerous criticisms, but its goal is to stimulate discussion and in turn shed some light on what tomorrow may bring.

The first question that should be asked is: what could trigger changes in water treatment technology? The most common answers are a change in potable water standards, or, more rarely, progress in related technologies, such as electronics, biotechnologies, membrane separation, polymer science, etc., especially spin-offs of developments in industrial water treatment or waste water treatment technology developed for specific purification purposes as ultrapure water. Additional reasons include the scarcity of water resources, the need for sophisticated reuse techniques, or the need for reducing costs. However, a sixth possibility, the opening up of new markets for water treatment and supply is almost never given as an answer. Moreover, the integration and relationship between all of these factors is usually overlooked. The result of traditional thought (the first three points) is an extrapolation of the status quo and a concentration of effort to improve and refine existing, conservative technologies rather than developing new ones. Let us examine in more detail some of the probable effects of changes in standards, technologies and markets.

It is not likely that water quality standards, which have become more stringent in recent years, will undergo major changes in the foreseeable future. Extensive risk assessment studies have been conducted world wide, the results of which indicate that marginal improvements in potable water quality are not justified in terms of the associated reduction in risk and increase in treatment costs. Hence, changes in treatment practices (even the simple use of granular activated carbon) will not be stimulated.

Emerging technologies, on the other hand, may bring about major changes in treatment practices. New adsorbants, better coagulants, and application of digital process controls are already on the horizon and promise appreciable progress. Membrane separation may change completely both treatment practices and the point set for acceptable quality of the water produced.

A more interesting, but also more controversial, force behind future trends in water treatment is the change in a market, the size and features of which are intimately related to changes in the available technology.

Water treatment is not generally regarded as an open market. The vast majority of water treatment plants are held by municipalities which do not act according to common market incentives or business rules. The evaluation of cost-effectiveness for a project is usually poorly made, the amortization period is abnormally long and willingness to take any risk in the design is limited. The restricted competition between both plant designers and manufacturers presents little incentive for change. The opportunity for change is very small. However as the industry becomes privatized this situation seems to be improving. Municipalities are contracting private firms not only to operate and maintain plants, but to design and construct them. This leads to more innovative designs. Included in this is the opening up of world markets as a result of a tremendous growth of the internal market of Japan, particularly in industrial and waste water treatment. The 1984 sales of water treatment equipment in Japan were twice those in the United States. This growth is coupled with ambitious research programmes which increase the likelihood that Japan, as it has done in so many other areas of technological competition, will seize a major share of the world water treatment market. We speculate, however, that this will stimulate competition, forcing other firms around the world to become more innovative in developing new treatment technologies.

This chapter is focused on the technical push which undoubtedly represents the drive of water treatment trends. Most of the techniques used in this industry are reviewed, emphasizing their possible future in the coming years.

2. AREAS OF POTENTIAL CHANGE

2.1 Clarification

Until recently, the approach to improving clarification has been to increase loading rates. From the first sedimentation basins with detention times of the order of days, there has been a trend towards higher and higher rated sedimentation units (e.g., plate and tube settlers, upflow and pulse clarifiers).

In addition, improvements have been made in the use of metal coagulants. It has been possible to modify their structure, transforming them into hydrated polymers which perform more efficiently (PBAC from $AlCl^3$ PIC from $FeCl^3$). These products are particularly useful for high loading or low temperatures. While these improvements have brought us to the point of near-perfect control over colloidal and suspended solid removal, removal of dissolved organic compounds by these processes remains mediocre at best. It may be that we have reached the kinetic limits of these processes. To surpass these limits we will need to improve our knowledge of fundamental mechanisms and/or employ new technologies.

Already finding use in flocculation, organic polymers promise further efficiency in coagulation-flocculation. Their ability to stengthen floc, increase floc density and decrease sludge volume has been well demonstrated. However, the problem of trace residuals of either monomers or polymers remains to be solved, perhaps by improved polymer synthesis or a better understanding of polymer–suspended solids–organic compound interaction. Also, with an improved knowledge of the flocculation mechanisms, it is possible to control the size of flocculated particles. This new possibility should help improve both the yield of the flocculation operation and the design of high flow rate separators. In the future there will be smaller, higher rated clarification units which may use a fixed media to remove floc from suspension, possibly making conventional settling obsolete? Contact or direct filtration is one such example. (There the trend has been to increase loading rates from some 2 m/day with slow sand filtration to rates as high as 20 m/h or more with direct filtration.) Just as the problem of filter clogging was solved in the past by deep bed, multi-media, and/ or constant rate filters, it is possible today to consider replacing transversal filtration with longitudinal flow filtration. Moreover, granular filter media may be replaced altogether in certain applications by filtration membranes. Such microfiltration is already in the planning stage for treating surface waters using a $0.2\,\mu m$ membrane.

The future of clarification will include radical transformations in suspended solids and organic removal, using direct membrane filtration without the addition of coagulants (the concept of the no-chemicals plant will be discussed in more detail later). In effect, it is foreseen that micro-filtration will be used to remove material in suspension, including a portion of the organic compounds found in the raw water. The remaining organic material will be removed using more specific techniques such as pervaporation alongside recently developed techniques such as ultrafiltration or reverse osmosis. The use of pervaporation will permit selective elimination of target substances by internal diffusion, creating concentrated, relatively pure wastes.

As far as reverse osmosis is concerned, new developments in membranes have already yielded a low operating pressure, thereby reducing energy costs associated with the application of this separation technique. Also, it is foreseen that new polymers will be used to provide ion selective membranes, applicable in special cases such as the rehabilitation of water resources containing a specific inorganic pollutant.

Not all of these techniques are currently feasible for application. While they now appear as technological novelties they will undoubtedly play an important role in the future of potable water treatment.

2.2 Oxidation, adsorption and disinfection

Ozone and chlorine have enjoyed a long history of use as oxidants and disinfectants. However, as analytical techniques have become more sophisticated, the problem of by-product formation (haloforms from chlorine, epoxides from ozone and chlorate from chlorine dioxide) has come to light. Since haloforms have been shown to be carcinogenic, the use of chlorine as a primary disinfectant has been reconsidered and research into alternative means of disinfection has been stimulated. The alternatives investigated to date either fail to provide a residual (ozone u.v. light), produce other potentially harmful by-products (chlorine dioxide), or are prohibitively expensive. In most cases oxidants are used without a fundamental knowledge of the mechanisms of their action. Present research is focussed on elucidating these mechanisms with the goal of pairing specific oxidants to specific waters and points of application. This will result in less by-product formation and perhaps increase the efficiency of disinfection against specific microorganisms.

Another method of removing organic compounds from water is to use adsorbents. This unit operation is already used routinely in drinking water production in Europe. However, the adsorbent used (activated carbon) is rather non-specific, which is good, but there is very frequently a fraction of the organic material remaining in the treated water. Improvements of the yield of the adsorption unit operation will be triggered by a better knowledge of the nature of the organic materials found in water. Research is now underway either to find new adsorbents or to combine an oxidation with the adsorption. One can foresee the development of new, tailored adsorbents specially designed to remove target organic compounds such as trihalomethanes. New activated carbon forms such as activated carbon fibres are now being produced in Japan. They may lead to new adsorption reactors.

The future of disinfection will lie in flexible, multi-point injection schemes, using combinations of disinfectants such as ozone H_2O_2 or H_2O_2 + silver. The possibility of solid, regenerable oxidants with specific functional groups would permit a high degree of oxidation-disinfection control. Organo-metallic compounds could be synthesized for specific treatment objectives. Such solid oxidants could be incorporated as media into a tangential flow filtration scheme, thus either physically removing pathogens or chemically inactivating them in one reactor.

Membrane filtration for sterilization has already found application in the food processing industry and holds some promise for water treatment. Membrane sterilization could be easily incorporated into the schemes for liquid–solid separation by membrane filtration described in connection with the 'no-chemical plant'.

2.3 Biological treatment

Future potable water treatment will make greater use of biological processes. Biological removal and transformation of nitrogen compounds are already commonplace. For more than twenty years nitrification of water containing up to 10 mg/l of ammonia has been accomplished by bacteria fixed on granular, aerated media. This process has undergone considerable inprovement, leading to the use of fine media and optimized aeration rates. The same sort of technology has been applied to iron removal and modified for nitrate removal. Such processes pose problems in terms of reliability as they require continuous monitoring. This factor would appear to make them unlikely candidates for 'point of use' applications.

Other possibilities which may be envisioned include *in situ* treatment of groundwaters. This has already been studied with respect to ammonia and nitrate removal but has met with limited success in past applications to iron removal. On the other hand, proven techniques such as electrodialysis and ion exchange may find future use in 'point of use' treatment. This concept, however, raises the problem of consumer safety with respect to the hazardous substances that may leak through a small unattended treatment unit. A first step to solving this problem is to sell a maintenance service along with a high technology treatment device. The next step is probably the design of fool proof treatment units including specific transducers connected to a monitoring unit. This device may shut down the treatment unit as soon as it detects malfunctioning parts, or untreatable inlet water. This monitoring unit may also tune the treatment device to produce constant predefined quality water. The following futuristic techniques could be used:

— utilization of membrane reactors which will allow both the fixation of specific species of bacteria and the elimination of biologically rate-limiting products;
— addition of specific enzymes to speed enzyme limited reactions;
— modification of bacterial species by genetic engineering.

While these last two processes will be technologically feasible by the year 2000, they will most likely be unacceptable from a cost viewpoint. Unless there is a major revolution in these technologies, they will remain applicable only for products sold at relatively high prices and having very simple composition, which is rarely the case in potable water treatment. They may however find an application in final polishing stages of water treatment.

2.4 Softening

The most common point of use treatment today is water softening by ion exchange on cationic resins. Larger scale softening requirements are met using chemical or catalytic precipitation. These two domains will be helped in the coming years by electric treatment. This treatment has been mastered and may be the basis of a new technology in softening and corrosion control applications, in conjunction with other softening treatments such as ion exchange resins; the protection of distribution systems from scale deposition and the protection of sulphate and chloride ion corrosion is also possible. Electrodialysis could experience a rebirth through the introduction of high-performance membranes.

2.5 Automation

It is within the domain of automation and digital control that the real revolutions in potable water treatment will occur. From now until the year 2000, the exponential growth of the microprocessor industry and the accompanying modifications in the human environment will be major factors in creating change through technology-push.

Timing relays were installed fifteen years ago in order to aid plant operation. Later, automatic control came into use in one or more stages of treatment using predetermined set points (Fig. 1a). Such systems were based on feedback schemes which minimized the chemical residual (aluminium, chlorine or ozone), presumably optimizing treatment. Kinetic considerations of each process were not explicitly considered. Today, the possibility of using mechanistic models in conjunction with microprocessors for real-time calculation of the status of a process allows a feed-forward control, with feedback information serving as a check on process operation (Fig. 1b). In addition, cost information on the operational characteristics of each process can be evaluated continuously, and process control altered accordingly to reach a cost-optimal treatment objective. This type of control has already been applied to ozonation. Further applications of this sort require: (1) a better fundamental knowledge of chemical and physical mechanisms involved in process operation, and (2) the development of reliable probes for controlling information on the state of chemical and physical parameters.

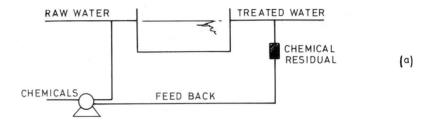

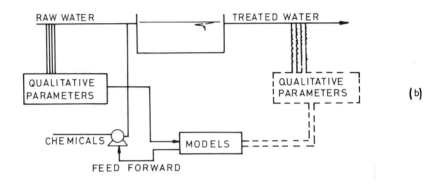

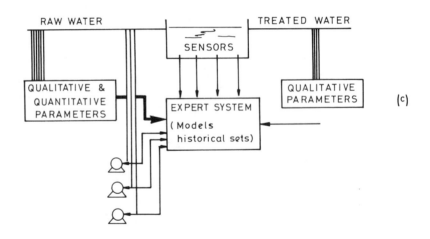

Fig. 1 — Trends in automation.

With the rapid evolution of microprocessors and programming capabilities, one can forecast for the near future 5th generation computers using high level languages capable of control based on past experience, in other words, artificial intelligence (Fig. 1c). In such a system, information on raw water quality will be related to process parameters and final water quality. The system will be capable of controlling an entire installation, correcting problems as they arise and learning from operational errors. Such an expert system will begin operation using rules taught to it by human operators.

Stochastic models may play a role, augmenting historic data and up-dating stochastic relationships accordingly, thus becoming more intelligent. We are not far from the time when a machine will control an entire treatment plant at optimum conditions, based on the data collected by the previous generation. One may imagine an advance order for water of a given quality and quantity based on forecast demands made possible by telecommand and process control using expert systems.

The no-chemical plant may be composed of a series of modular components, each consisting of a semi-permeable membrane, specific for the removal of a particular contaminant. The only input into water treatment would be energy to maintain sufficient pressure and for membrane cleaning. Individual modules might be activated on command to provide water tailored to the needs of specific consumers.

This type of treatment scheme might be suitable for point of use applications. The home of the year 2000 will possess its own treatment plant, controlled by the home computer, delivering water with characteristics specified by the consumer. But enough dreaming!

It is, however, certain that the cost of not considering the future of the water treatment industry in such fantastic terms is to condemn it to be an outdated, inefficient industry.

3. CONCLUSION

Changes in water treatment technology have been deceptively slow during the last several decades. It seems, however, that the time has come for major changes. Countries with advanced technology, like Japan, will set the pace. This does not mean that there will be no room for the so-called adapted technologies, in particular in the developing countries. It merely implies that the general trend towards change which will also influence the developing countries comes from the combination of a new market-pulling and a strong technology-pushing in the countries with strong economies. The new technologies that we have been talking about will probably yield water treatment plants featuring high energy yields and the absence of the need for maintenance and reagents. This revolution may in turn trigger a new market-pull as developing countries are likely to open their market to such a technology.

The world is changing even in water treatment.

3.2

Emerging technologies in the water supply industry

J. F. Manwaring,
Executive Director, AWWA Research Foundation, 6666 W Quincy Avenue, Denver, Colorado 80235, USA.

1. INTRODUCTION

The AWWA Research Foundation, acting on behalf of the water supply industry of North America and under a cooperative funding agreement with the US Environmental Protection Agency, designed and administered an 'Emerging Technologies Project' in the summer and autumn of 1984. The collaborative effort brought together representatives from eleven countries — Belgium, Canada, France, Italy, The Netherlands, Norway, Spain, Switzerland, The United Kingdom, The United States and West Germany — in a three-day seminar to discuss 161 individual projects which the participants felt had some applicability to the theme of the conference.

The objective of the project was to identify emerging technologies — that is, innovative developments in all aspects of the water supply field — with two purposes in mind: the collection and exchange of information among the participating parties; and the discovery of promising approaches that merit further research, development and/or application. In addition it was anticipated that the projects would lead to cooperative ventures and continued information exchange among the participants.

The AWWA Research Foundation's effort differs from the 'Environmental Technology Workshop' at the University of Cambridge in several respects. First, the Research Foundation's project concentrated solely on drinking water, but covered all aspects of the topic, including resources, planning, treatment, distribution and management as opposed to focusing upon technology and micro-organic contaminant removal. The design of the Foundation's seminar did not identify any priority topic but several issues did emerge as common concerns and topics. In many ways the 'Emerging Technologies' effort was an unintentional complement to the 'Environmental Technology Workshop' by providing an examination of the current technological direction of the water supply industry along many different operational fronts.

It also afforded an opportunity to draw some comparisons between the research and directional priorities of the drinking water industry in North America and Western Europe.

The AWWA Research Foundation has always been interested in participating in joint research programmes where the specific project has equal participation by, and benefit for, the involved parties. These activities range from direct research efforts, such as the taste and odour study presently being sponsored by the AWWARF and the Société Lyonnaise des Eaux, to the preparation of state-of-the-art reports. Examples of the latter include the AWWARF/KIWA report on the removal of organic contaminants and the AWWARF/EBI report on the internal corrosion of water systems. The 'Emerging Technologies' effort was another example of a cooperative and collaborative effort.

The following report attempts to summarize the major conclusions to be drawn from the Foundation's effort and to show contrasts among the various approaches used by the participating countries. Thus these two efforts — 'Emerging Technologies' and 'Environmental Technology' — provide a perfect example as to how independently designed and executed programmes can complement and supplement one another without reducing the effectiveness of either.

2. OVERVIEW

The Amsterdam meeting of autumn 1984 was preceded by several months of report preparation, review, and revision by the attendees. The 161 projects were documented, compiled and distributed via workbooks to the 15 participants prior to the conference. The project information included a description, development status, operational and cost data and place of installation.

Table 1 shows the individual titles of the projects as submitted by each country, under the five categorial headings. A reader who wishes to pursue any particular project may contact the Research Foundation for further information. It must be emphasized that this body of research is a representative sample of the combined countries' efforts, not an all-inclusive collection. Figures 1 and 2 provide a summary of the information submitted during the 'Emerging Technologies' project. As shown in Fig. 1, almost 60% of the projects were within the Treatment and Operations category, indicating a definite preference for research in this area. Figure 2 shows a further categorization of the projects into eight major topical areas; corrosion may include projects from the treatment, distribution, and water quality categories; the same is true of other topics listed.

In general the new successes with more expensive technologies grabbed the spotlight; however optimization of existing unit processes emerged as the guiding, practical theme, driven by mutual concern for water quality, operational costs and the national economies. The majority of optimization projects focused on modifications to improve conventional treatment efficiencies, remove source contaminants and reduce treatment by-products.

It was a consensus of the group that growing demands and limited water resources, in an increasingly contaminated and populated environment, will require not only reduced consumptive patterns and utility loss control programmes but also responsible and resourceful management by the contributors of hazardous and

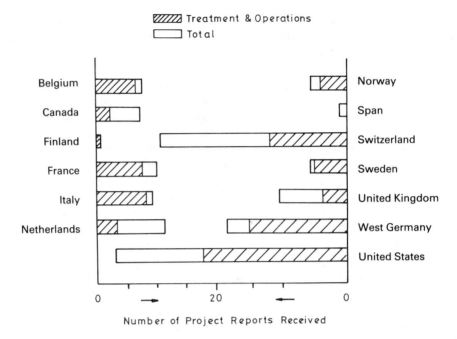

Fig. 1 — Comparison of project submissions.

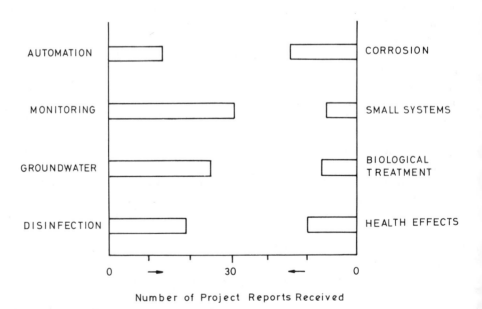

Fig. 2 — Major topical areas.

deleterious waste waters. This perspective merely reinforced the knowledge that water supply resources are intricately tied into other environmental control programmes and that comprehensive planning necessarily considers all aspects of the environment.

Another general observation was that the European water supply community is directing its attention towards a broad spectrum of micro-organic contamination as opposed to the American focus which seems to be on individual organic compounds. For instance, the United States has a great deal of research directed at the removal of specific contaminants such as trichloroethylenes, trihalomethanes, etc., whereas the European countries seem to be investigating technologies directed at the organic problem as a whole. Bank infiltration, reinfiltration, and in-ground storage are just a few examples of European technology designed for the wide spectrum of organic contaminants.

3. Automation

Automation of water supply operations, continuous monitoring, and modelling are being pursued with vigour by researchers in all countries to resolve diverse issues in management, operations, analyses and watershed surveillance. Over forty projects reflected these approaches. Primary examples were:

— the Swiss computerized management model and monitoring network for expansion of Zurich/Hardof groundwater production which involves artificial recharge, groundwater table fluctuation and water quality control;
— the French and United States plants' centrally controlled computer for management, operations, and in-line micro-hydropower recovery;
— the British Chertsey project which utilized centrally controlled microprocessors to replace ten plant operators; and,
— Madrid's automatic distribution system monitoring network for water quality with computerized feedback for operational response.

Emerging from discussion of these automation projects was a challenge to develop more software, both compatible and flexible, to address the practically open frontier of applications on available hardware. Also, the reviewers sought equitable and cost-effective approaches for management to optimally structure jobs which balance qualified personnel and their workloads with automated operations.

4. Disinfection

Ten years ago Europe's experience with ozone and chlorine dioxide matched the American's research needs for alternative disinfectants. However, the utilities in the United States continue to cautiously avoid major shifts towards either of these disinfectants, choosing instead chloramines as an alternative to free chlorine. Ozone is the preferred and established practice in most European countries which have avoided chlorine, historically for its taste, and currently for its organic by-products. Current European technology applications for ozone reflect this preference. Belgium, France, the United Kingdom, the Netherlands and Switzerland have much time and funding invested in ozone disinfection including projects involved in photochemical generation, u.v. monitoring of ozone levels, and development of new reactors and applications.

In addition to its disinfecting capability, ozone is being studied on both sides of the Atlantic as a pre-oxidant and as a means to enhance flocculation and filtration effectiveness. The most notable United States facility examining all integrated aspects of ozone application is the 3 million gallons per day (mgd) pilot plant in New York City.

Several countries are working to eliminate the chlorite problem inherent in the production of chlorine dioxide, including a German device to produce pure chlorine dioxide and a Belgian continuous monitor for residual chlorite. The Italians are evaluating hydrogen peroxide for dechlorination and deozonation. Positive u.v. irradiation results were offered by the Swiss for systems requiring disinfection of large flow rates (400–10 000 m^3/h) at a cost of 1 cent per m^3 for 25 mWs/cm^2 dosage.

There has been, and continues to be, a great deal of work in the United States on trihalomethanes — specifically the reduction of THMs while maintaining the bacteriological integrity of the distribution system. The common approach has been the continued use of chlorine as the primary disinfectant but in a manner which minimizes the production of THMs, for example using combined chlorine, changing the point of chlorination and improving flocculation are just a few of the methods being used at operational treatment plants. The European community, because of its historic reluctance to use chlorine is not experiencing the THM problem common to American water systems.

5. MONITORING

Thirty projects reflected pronounced emphasis on up-dating the monitoring of source, unit process and distribution system water qualities. Continuous analytical systems for most water quality parameters and treatment efficiencies and rapid detection methods for bacteria were reviewed. These techniques are available for controlling and optimizing water systems. More field testing and routine applications on a full-scale basis are obvious needs.

Over the past 25 years the Europeans have developed and used rather sophisticated fish sensors to monitor the quality of their surface water supply and control their intakes accordingly. System variables among the British, Swiss, Dutch and Canadian projects included: simultaneous water quality analyses, fish species and function to monitor, potability, water sampling, alarm and control strategies.

Mutagenic testing of water (both raw and finished drinking water) appears to be a widespread practice in Europe, at least in comparison with the United States and Canada. Mutagenic monitoring in France, the Netherlands and the United Kingdom has been on-going since the late 1970s and early 1980s. Data are being collected as background information and as a gross measure of contamination. While all agree that there is no meaningful interpretation of single results, mutagenicity is being used by the European community as an indicator signalling the need for more detailed follow-up and analysis.

6. BIOLOGICAL TREATMENT

While the Europeans have embraced biological treatment, particularly for microorganics and nitrates, the process has been deliberately neglected by the American research community. Operational denitrification facilities in the United Kingdom, France and Germany appear to be both efficient and effective. The Swiss use *in-situ* biological denitrification while the Germans are experimenting with the decontamination of groundwater by treatment with ozone to enhance biological degradation and reinjection. Bank infiltration which utilizes the natural biological purification process has been practised in West Germany and the Netherlands for several years; the utility of the process is now being investigated at New Orleans and Louisville in the United States.

The major objections to biological denitrification in the United States emanate from the regulatory community and are based on two factors. First, any such system must be designed to provide a growth environment for bacteria which unfortunately includes opportunistic pathogens. Second, it is necessary to add an organic food source to what may be an organics-free water supply.

7. IRON AND MANGANESE

Water quality problems associated with iron and manganese in groundwater appear to be ubiquitous and different techniques are being investigated. In the Netherlands and Sweden for instance, as well as on Long Island in the United States, a process utilizing injected water and/or air is being employed to generate *in-situ* oxidation of iron and manganese in the aquifer. Norway is investigating the use of reinfiltration as a natural means of removing these elements. Studies in the United States and Canada are focusing on the simultaneous addition of sodium silicate and chlorine to sequester the iron and manganese.

8. SMALL SYSTEMS

Like corrosion and iron/manganese, the water quality and operational problems associated with small water supply systems are a world-wide concern. The primary objective of small system research can be stated as the development of the technologies that have low capital costs, low chemical usage, low operational costs, minimal operational requirements, and low sludge production. As impossible as it would appear to simultaneously meet these objectives, many countries are investing a great deal of money in an effort to do so.

Norway, for example, is researching the use of reverse osmosis and electrocoagulation for small systems applications. Sweden is examining contact filtration and alkaline media filtration, while Switzerland is researching the use of horizontal roughing filters. The United States is re-examining the operational requirements and costs of an old technology, slow sand filtration, for its applicability to small system situations.

9. CORROSION

Every country, without exception, is investing heavily in the control and resolution of corrosion. Several researchers are involved in attempting to develop a better surrogate parameter than the current indices, while some are focusing on the measurement and reduction of specific corrosion by-products. All agree that corrosion is probably the most pervasive economic problem facing the water supply industry today. However, it is an extremely complex phenomenon that requires site-specific investigation; research is just beginning to unravel the many competing and synergistic factors involved in corrosion.

10. REGROWTH

One of the emerging problems mentioned by several representatives was bacterial regrowth in water distribution systems. Belgium and the Netherlands are just beginning to study the cause of such regrowth and factors which control it. Outbreaks of regrowth in several United States water systems have caused a re-examination of the basic premise of post-treatment disinfection. It appears to be one of the common areas of future exploration.

11. CONCLUSIONS

Each country has its own established drinking water organizations with research agendas and priorities developed over the years. Although no specific cooperative venture among the participants was developed during the conference, all gave enthusiastic support to continuing this activity and considered the compilation and review of projects a valuable resource and an excellent centre point for further contacts. It was felt that research institutions must actively communicate the need for, and value of, research to the industry and the public.

An additional purpose of the 'Emerging Technologies' project was to enrich and enlighten the AWWA Research Foundation planning process. Therefore, members of the Foundation's Research Planning Committee were able to use the information uncovered during the conference to assist in examing and selecting the Foundation's own research projects for 1985.

Reflecting on the experience with this project, there are two observations to describe. The first concerns what is happening in the United States. Without doubt, the past ten years or so have witnessed a remarkable new awareness of the need for research and development. American water suppliers, by and large, are giving genuine assent to the notion that there is much to be learned, that they cannot rest their case on the conventional treatment of thirty years ago, and, indeed, that it may be possible for research someday to provide relief to the perennial problems that plague the industry. These may seem to be self-evident truths, but they have not been so among the rank and file of the 60 000 water suppliers in the United States. Proof that the new attitude towards research is not an illusion of a hopeful researcher is the number of dollars and the level of effort being invested in research by more and more utilities. The increasing interest and support for the AWWA Research Foundation may be taken as a measure of the conviction that the industry has to unite in a research effort of respectable proportions.

The second observation is that water researchers in the developed nations of the world must cooperate with each other more extensively — more openly — than they have in the past. There is no benefit in withholding information for reasons of national chauvinism or professional pride. Copyrights and patents are to be respected, of course, and credit should be lavished upon those who advance the science and technology. Information exchange, however, and collaborative efforts should become habits of mind for those whose scientific and technological work is in the service of the people.

Table 1 — AWWA Research Foundation. Research projects reviewed at Emerging Technologies Conference, Amsterdam, The Netherlands, 16–18 October, 1984

Water resources
Canada
— Water supply reliability and risk.
The Netherlands
— Recharge wells.
Switzerland
— The most up-to-date ground water management by means of process computers explained by the example of the ground water plant Hardhof, Zurich.
— Progressive water tariff as incentive to water saving.
— Supplying water to small communities and entities from a central regional water production facility.
— Tariff measures and operational possibilities for the restriction of specific water consumption.
United Kingdom
— Intake protection systems.
— Water well development/rehabilitation.
— Catchment quality control.
United States
— In-reservoir chlorophyll monitoring as a reservoir management tool.
— Multiple water supply approach for urban water management and alternative technologies for small water system management.
— Water audit guidelines for conservation and management.
— Laser mapping.
West Germany
— Ground water enrichment with pretreated surface water.

Water treatment and operations

Belgium
— Package treatment for purifying water heavily contaminated by nuclear, biological or chemical agents.
— Development of a continuously operating analyser for monitoring residual chloride in water.
— Photochemical generation of ozone.
— Optmization of activated silica preparation.
— Optimization of chlorine dioxide generation for post-disinfection.
— Preozonation as an aid in flocculation-filtration.

Canda
— Sequestering of iron and manganese; treatment of contaminated ground water.
— Reducing trihalomethanes in finished water.
— Organics removal by conventional treatment, add-on activated carbon treatment and aeration.

Finland
— Biological and chemical removal of iron and manganese from groundwater (larger systems).

France
— Monitoring and ozonation process through u.v. measurement.
— Development of a new ozonation reactor: the deep U tube.
— Biological aerated filters or biocarbon®.
— Chromium removal from ground water.
— Biological denitrification of ground water.
— Use of prepolymerized Al-OH solutions as primary coagulant/flocculant.
— Compuertised control and total automation for a drinking water treatment plant of 180 000 m^3 per day.

Italy
— The use of solar energy to power remote pumping stations.
— Water reuse through the rim-nut process.
— Removal of organic halocompounds in drinking water by (aeration), air stripping and activated carbon (GAC).
— Anaerobic treatment of concentrated wastewaters.
— Advanced precipitation processes for heavy metals removal from wastewaters.
— Deozonation with hydrogen peroxide.
— Dechlorination with hydrogen peroxide.

The Netherlands
— Side-effects of past chlorination.
— Removal of methane with aeration.
— Removal of volatile organic substance by aeration.
— Undergound iron removal from ground water.

Norway
— Humic substance removal by ion exchange.
— Humic substance removal by reverse osmosis.
— Electrocoagulation for removal of aquatic humus from drinking water.

— Guidelines for the planning, construction and operation of submarine pipelines for water supply and sewage.

Sweden
— Contact. Filtration using AIB filters.
— Alkaline media filter for installation in the pipe system (small systems).
— Purac's FLOOFILTER®.
— Dynasand continuous sand filter.
— Vyredox, *In situ* purfication of ground water.

Switzerland
— Simple and compact process unit for iron removal from anaerobic ground waters.
— Treatment of Karstic spring waters containing chlorinated hydrocarbons.
— u.v.-water disinfection for large flow rates.
— Oxidation of organic matters using u.v. in conjunction the hydrogen peroxide.
— In-ground biological denitrification of ground water.
— Production and application of high ozone concentrations in water treatment plants.
— Covering the surfaces of enrichment basins and slow filters with Fleece-Mats.
— The effectiveness of rapidly operated slow filters.
— Pre-oxidation of surface water and bank infiltrate with a mixture of chlorine and chlorine dioxide.
— Return of treated flume water to the raw water.
— Elimination of trace organic compounds by infiltration of river water into ground water.
— Horizontal roughing filtration as pretreatment for slow-sand filters in developing countries.

United Kingdom
— Chertsey automation project.
— Removal of volatile organics by aeration.
— Studies on use of ozone in water treatment.

United States
— Air wheel drive for flocculation equipment.
— Ozone pretreatment: effects on biologically activated carbon, disinfection and treatment by-products.
— Low head filter backwash design.
— Coagulant control test apparatus.
— TCE removal from ground water using aeration in Smyrna, Delaware.
— Aquatic plant pilot and bank filtration project, New Orleans, Louisiana.
— Pretreatment of water using granular activated carbon.
— Effective filtration methods for small water supplies.
— Ultrafiltration of surface water for colour, TOC and THMFP reduction.
— Nutrient film technique for wastewater renovation.
— Granular activated carbon as a barrier against contamination.
— Trident® water systems.
— Ulta-sensitive electronic turbidimeters.
— Automated jar testing system: optical floc testing by microcomputer.

- Closed-loop stripping analysis for determining taste-and-odour causing compounds.
- Cost effective optimization of filtration plant performance utilizing new technology.
- Water treatment with activated oxygen (Photozone).
- Surface-wash systems for filters using the Baylis Nozzle.
- Radium selective complexer for radium removal from potable water.
- Ann Arbor controls trihalomethane.
- Reduction of total trihalomethanes by alternatives treatment methods.
- Hydroperm cross flow microfiltration.

West Germany
- Activated carbon adsorption for removing chlorinated hydrocarbons from ground water.
- Removal of volatile halogenated hydrocarbons by air stripping and activated carbon adsorption.
- Energy-input-controlled direct filtration (Wahbach system).
- Specially designed approaches for the production of pure ClO_2 solutions using chlorine.
- Compact high affinity flocculation plant (CFP).
- Production of highly purified aqueous calcium hydroxide solutions.
- Powdered carbon filtration.
- Refiltration flocculation ('REFIFLOC' Process).
- Upflow filtration for nitrification, denitrification, and iron and manganese removal.
- Treatment of ground waters contaminated with volatile organic substances.
- Macroreticular ion exchange and biological treatment of a reduced coloured ground water.
- Reinfiltration of ground water after ozone treatment and oxygen enrichment.
- Aeration using corrugated fibre sheets
- Two-stage fluidized bed incinerator for reactivation of GAC.
- Advanced wastewater treatment for ground water recharge.
- 'The Mulheim Process' for treating river waters using ozonation and biological treatment in GAC filters.

Water quality

Canada
- Continuous monitoring of raw water for toxic spills.

France
- Quantitative analysis of health-related organics at low concentration levels.
- Analysis of organics and their mutagenic activity in drinking water treatment.

Italy
- Monitoring raw water quality parameters.

The Netherlands
- Improvement of water quality during storage in reservoirs.
- Automated isolation and sample preparation for the chemical and toxicological analysis of organics.

- Organic halogen determination (OCl, OBr and OI).
- The modified fouling Index as a measure of the fouling potential of injection well water.
- Detection and evaluation of mutagenic activity in drinking water.
- Easily assimilable organic carbon (AOC) in drinking water.

Spain
- Localized aeration as a means to break reservoir stratification around an intake tower.

Switzerland
- Continuous monitoring of several quality parameters of a river's water.
- Electronic system to monitor the effects of drinking water pollution and toxicology on trout.
- The continuous control of quality parameters during the treatment process of drinking water.
- Continuous spectrophotometric measurement of chlorine dioxide and chlorite using χ-tolidin.
- The utilization of fish-test facilities for the control of water quality.

United Kingdom
- Mutagenicity testing facilities.
- Determination of assimilable organic carbon in water.
- Rapid detection of bacteria in water using impedance measurement.
- Microbial tracing of water pollution.
- Non-volatile organics in drinking water.
- Biological screening tests for toxicity.

United States
- Isolation and identification of non-coliform bacterium from potable water.
- Super sensitive luminescent bacterial bioassay for assessing toxicity in potable water.
- Flavour-profile analysis of drinking water.
- Immunoflurescence for detection of *Giardia lamblia* Cysts in drinking water.
- Rapid bacteria detection instrument.

Distribution systems

Belgium
- Identification of bacteria which cause aftergrowth in water mains.

Canada
- Evaluation of alkalinity as a predictor for corrosive activity.

France
- Optimal control of a large water supply network: West Paris case study.

The Netherlands
- Conditions of tap water.

Norway
- Water treatment for corrosion control using lime and carbon dioxide gas.
- Lime dosing system for corrosion control purposes.

Spain
— Quality monitoring in the distribution system: feed-back by computerized analysis.

Sweden
— Water treatment for corrosion control using calcium carbonate and hydrochloric acid.

Switzerland
— REKA-coupling for fibre-cement pressure pipe
— Automatic control of large water distribution networks with decentralized treatment facilities.
— Internal refurbishment of a large-calibre tamped concrete duct.
— Internal refurbishment of lead sleeve connections on cast-iron pipes.
— pressure reliability in drinking-water ducts.

United Kingdom
— Power recovery in water distribution systems.
— Digital recording of water mains and associated information.

United States
— Realistic replacement/rehabilitation criteria for distribution system components.
— Flow monitoring utilizing ultrasonic techniques and dye injection.
— Analysis of water main breaks.
— Retrofitting small hydroelectric generating plants into an exisiting water distribution system.
— Hydraulic simulation model.

West Germany
— Use of a fiberscope for control purposes in the Berlin distribution system.
— Special cement mortar lining of ductile cast iron pipes.

Miscellaneous

Canada
— Micro-computer application to water supply technology.

Switzerland
— Conception of an emergency water supply.
— Distribution and storage of drinking water in plastic bags for emergencies.

United Kingdom
— Protective measures against the nuclear destruction of electrical equipment by electro-magnetic radiation.
— Effects of effluent recharge on ground water quality.
— Simulation of ground water quality, with particular reference to nitrate.

United States
— Hydrogenerating units in water supply systems.

3.3

Trends in water treatment technology: disinfection, oxidation and adsorption

V. L. Snoeyink
Professor of Environmental Engineering, Department of Civil Engineering,
University of Illinois-UC, 205 North Mathews, Urbana, Illinois, 61801, USA

1. INTRODUCTION

The presence of different types of organic compounds in United States drinking water supplies has been a major force for change during the past decade and promises to continue to be so in the near future. The establishment of a standard for total trihalomethanes (TTHMs) has led to significant change in chemical oxidation and disinfection practice, and this change in turn has had an impact on the effectiveness of the barriers to transmission of disease-causing organisms and to growth of bacteria in distribution systems. It is also leading to greater use of ozone in the United States.

The widespread occurrence of volatile organic chemicals (VOCs) in groundwater has led to more frequent application of organic compound removal technology. Publication of various incidences of groundwater contamination has led to increased public concern about poor water quality, and this may eventually lead to the increased use of organic compound removal processes for our surface water supplies.

2. DISINFECTION AND CHEMICAL OXIDATION PRACTICE
2.1 Change to monochloramine

The TTHM standard of 100 μg/l has impacted both those utilities that produced water which exceeded the standard and had to change their treatment to meet it, and those that did not exceed the standard but still modified treatment to minimize TTHM. A common approach to achieve this reduction has been to change from free chlorine to monochloramine; in addition to those that have changed, it should be noted that a large number of communities throughout the United States have used monochloramine for more than twenty years [1]. The change can be made in a number of ways, including:

— adding chlorine to water containing ammonia at a dose much less than the breakpoint dose (i.e. <5:1 molar ratio, chlorine to ammonia);
— first adding ammonia to water that contains insufficient ammonia, then chlorine at a dose less than the breakpoint dose;
— adding sufficient chlorine within the treatment plant to achieve a free chlorine residual, followed after some time by addition of ammonia to convert the free chlorine to monochloramine;
— addition of an alternative oxidant such as ozone or chlorine dioxide within the plant, followed by application of monochloramine as the water leaves the plant in accordance with the first two steps above.

Significant differences result depending on which of these options is used. There are differences, for example, in the concentrations of TTHMs formed, in disinfection efficiency, in the effectiveness of chemical oxidation, and in the biological stability of the water, both within the plant and in the distribution system.

Addition of chlorine to water containing ammonia, which was purposely added or which was present from natural sources, appears to have the greatest impact. Figure 1 shows very little TTHM is formed by 5.5 mg/l NH_2Cl as Cl_2 in comparison

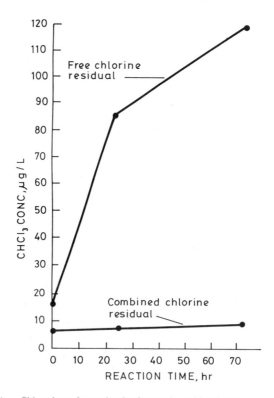

Fig. 1 — Chloroform formation by free and combined chlorine residual [2].

with the same concentration of free chlorine [2]. If this treatment is to be used, however, it must be kept in mind that oxidation and disinfection will be less extensive. Monochloramine is not able to oxidize manganese (+II) for example, whereas free chlorine above pH 8 can oxidize it. Monochloramine is also not able to oxidize colour as effectively as free chlorine and may necessitate the use of larger coagulant doses to remove the colour. Data presented by the National Academy of Sciences Safe Drinking Water Committee [3] were used to develop the comparison shown in Table 1. As shown, free chlorine is a much more effective bactericide, but

Table 1 — NH_2Cl disinfection efficiency

		Concentration (mg/l)	Time (99% kill, min)
E. coli	Free Cl	1	0.5
	NH_2Cl	1	175
		2	88
Polio virus	Free Cl	1	5
		0.1	50
	NH_2Cl	2	450
		18	50

by using higher concentrations of monochloramine and longer contact times, adequate kills can be achieved. Indeed, lower microorganism counts are often observed at the ends of distribution systems after a change from free chlorine to monochloramine because the slower-acting monochloramine is able to persist to the end of the system [4]. The virucidal effectiveness of monochloramine is much less than that for free chlorine, however. As shown in Table 1, a monochloramine concentration of 2 mg/l as Cl_2 requires 450 min to achieve a 99% kill of polio virus. Thus, changing from free chlorine to monochloramine significantly reduces the barrier to passage of virus through the treatment plant and distribution system to the consumer; if there is a danger of virus contamination of the water supply it appears that the sole use of monochloramine as a disinfectant should be avoided.

A major advantage to using monochloramine is considered to be cost. Ammonia is added to water in a manner similar to chlorine and thus presents no significant operational problems. Symons et al. [5] report that the total cost of monochloramine doses of 5 mg/l as Cl_2, or less, is less than 1 ₵/m³ (4₵/1000 gallon). If sufficient ammonia is present in the water, the cost is lower by a factor of 2–3. Indeed, a change from free chlorine to monochloramine may reduce costs because a lower chlorine dose may be required. Other alternatives are much more costly (see below). However, if additional costs are incurred, such as for increased coagulant to remove colour, for increased permanganate or other oxidant dose to remove iron, manganese or odour, or for an alternative disinfectant to provide a substantial barrier to virus passage, the total cost of changing to monochloramine will be much higher.

An improved barrier to virus is possible if free chlorine is maintained for a time before ammonia is added to produce the combined residual. A free chlorine concentration of 1 mg/l for only 5 min will produce a 99% kill of polio virus, for example. Of course, more TTHM will form, but the trade-off of some TTHM for an improved barrier to virus seems reasonable for some waters. The state of Kansas requires that all supplies use monochloramine, but also that a free residual of at least 1 mg/l be maintained for 30 min before ammonia is added [4]. The chlorine demand to produce a free chlorine residual will depend upon the concentration of ammonia and other reduced substances in the water. Ammonia concentration is especially important, because if it is more than a few tenths of a mg/l, the breakpoint dose will probably cause formation of undesirable concentrations of TTHMs. Oxidation of the reduced substances using a biological process before chlorine addition seems particularly appropriate. Use of ozone or chlorine dioxide instead of free chlorine would eliminate the TTHMs that form during that time, as discussed below.

2.2 Biological stability

Consideration must be given to the biological stability of the water if monochloramine is to be used as the residual disinfectant in the distribution system. Although many communities use monochloramine without reported problems, situations have been documented in which a monochloramine residual has not been sufficient to prevent extensive growth of microorganisms in the distribution system and its attendant poor water quality problems [6]. Extensive growths can promote corrosion and red water problems, odour and other aesthetic problems, regrowth of coliform species such as *Klebsiella pneumonia*, and other difficulties, so consumer complaints are usually abundant when these growths are present. The specific conditions that result in the development of microbial growths in distribution systems have not been clearly defined. Certainly, a microbial substrate such as biodegradable organic matter or ammonia is necessary. The use of a weaker disinfectant, such as monochloramine in place of free chlorine, may be a factor as may the presence of tubercules, sediment and other niches in which the organisms may grow in a protected environment. Certainly higher concentrations of monochloramine will be required to prevent growth than was necessary for free chlorine if the growth occurs in tubercules, etc., in contact with water containing chlorine. More research is needed, so that we can better predict what water quality will result in growths, but I am of the opinion that the switch to monochloramine as the distribution system residual will make it more important to make the water biologically stable before distribution.

Biological stability can be achieved by using a biological treatment process as part of the treatment train. Traditional United States practice has been to take extensive measures to prevent biological growth within the treatment plant. Chlorination before flocculation and sedimentation basins, sand filters and activated carbon filters has been used to prevent algae and bacterial growth, whereas in other countries, sand and activated carbon filters have been allowed to double as fixed-film biological reactors that remove biological substrates [6]. A change in chlorination practice to eliminate pre-chlorination should permit more biological growth on filter media and thus produce a more biologically stable water if sufficient oxygen is present, but operational problems may develop because: (1) algae growth in flocculation and sedimentation basins and/or extensive biological growth on filters with relatively fine media (0.5 mm effective size sand is very common in the United States) may lead to short filter runs; (2) the backwash system for our filters may not provide sufficient scouring to remove excess growth from the sand; and (3) rotifers, nematodes, etc., may develop within biologically active activated carbon filters, and these may penetrate the filter and appear in the distribution system. Alternatives which would minimize these problems include placing a biological, fixed-film process first in series in the treatment train [6]. Any organisms which would slough off could then be removed by subsequent coagulation, sedimentation and filtration. Also, larger diameter media in deeper beds with air scour could be used in place of the existing rapid filters to avoid operating problems caused by microorganism growth.

Biological oxidation can be carried out in packed beds or upflow fluidized beds. The oxygen demand of the water will determine whether or not aeration prior to biological oxidation will be sufficient, or whether additional air must be added to the process. Assuming supplemental air is not required, Short [7] showed that biological oxidation to remove ammonia was cheaper than breakpoint chlorination if the ammonia concentration was greater than ca. 0.7 mg/l. The low cost makes it reasonable to consider using biological treatment before chlorine is added; the result should be a water with fewer chlorinated organic compounds (because lower chlorine doses are possible) and which is easier to distribute. An important unknown which remains is how much, if any, biodegradable organic matter is removed in conjunction with the ammonia.

2.3 Ozone

Ozone is an excellent bactericide, virucide and chemical oxidant that is used extensively in Europe but which has not been applied very much in the United States. There is some indication that the situation is changing, possibly because of the chlorinated organics that result from excessive use of chlorine and the benefits of ozone. An informal count in 1984 showed that more than twenty plants were in use, under construction or under design, compared with two plants that were in use in 1977. The reasons for the increased usage include TTHM reduction, disinfection, Fe and Mn oxidation, colour removal and use as a coagulation and filtration aid. While the disinfection and colour removal properties of ozone have been known for some time, it is only recently that the ability of ozone to aid coagulation and filtration has been recognized. Typical results showing the effect of pre-ozonation on the quality of effluent from a direct filtration plant are shown in Fig. 2. These data were obtained from a pilot study for the Los Angeles Department of Water and Power, and they show a significantly better effluent quality when ozone was used than was obtained when chlorine was used as the pre-disinfectant or when no pre-disinfectant was used. Similar results were reported by Fiessinger *et al.* [8]. This effect is dependent on water quality, however, because at the Los Angeles pilot plant, no difference in filter effluent quality was noted between pre-ozonation and pre-chlorination when the raw water was changed to a higher ratio of groundwater to surface water. Much more needs to be determined about the factors which cause ozone to work in this manner.

The effect of pre-ozonation on the performance of granular activated carbon (GAC) columns is also worthy of note. Ozone can react with certain compounds and the reaction product is more easily degraded, although frequently it is also less adsorbable. Passage of pre-ozonated water through a biologically active GAC column may show additional removal by biological oxidation. Glaze *et al.* [9] studied

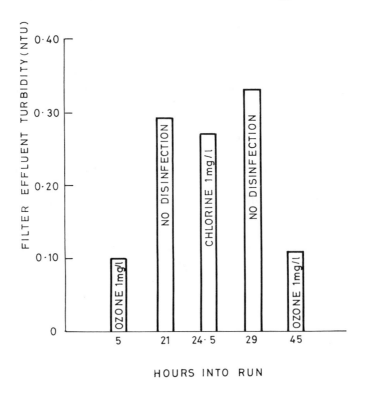

Fig. 2 — Effect of pre-treatment on filtered water turbidity (21 October 1978, influent turbidity 1.5–1.8 TU).

this effect at Shreveport, Louisiana, USA, using ozone doses of 2–3.5 mg/l for a water containing 6–12 mg/l TOC. They found evidence of a slight increase in the life of the carbon for removal of trihalomethane precursors (*ca.* 10%) and concluded that optimization of the dose may lead to greater increase in carbon life. They also note that this effect is likely to be an important function of the type of organic matter in the water. The above effects, coupled with the desire to reduce TTHM without sacrificing disinfection efficiency, should lead to much greater use of ozone in the United States.

Problems can result from the use of ozone. There has been ample demonstration that ozone can increase the concentration of biodegradable organic matter through partial oxidation of non-degradable compounds. For waters in which this occurs, it is important to follow the ozonation process with biological oxidation to avoid the problem of increased bacterial growth in the distribution system, or to avoid having to use too large a concentration of residual disinfectant to control the growth. A further problem is our incomplete knowledge of the chemistry of ozone. We do not have the necessary knowledge of the kinds of end-products that can form under various conditions and it is most important that we undertake research to correct this situation. There is a similar lack of knowledge of chlorination end-products, but because of the proven worth of chlorine as a disinfectant, there is a natural concern that we should not be too quick to replace it with something that may prove to form just as many questionable endproducts.

The cost of using ozone was reported by Symons *et al.* [5] in 1981 to range from *ca.* 0.6 c/m^3 (2.4 c/1000 gallon) for a 1 mg/l dose to 1.2 c/m^3 (4.8 c/1000 gallon) for a 5 mg/l dose for a $37\,800 \text{ m}^3/\text{d}$ (10 MGD) plant. This is considerably more expensive than chlorine, but the cost may be offset by the benefits of ozone compared with chlorine or monochloramine. For example, if ozone is used before GAC, a decrease in regeneration frequency from 2 to 2.8 months can offset the cost of adding 2 mg/l of ozone to the water. Although a decrease of this magnitude is not likely, some decrease is expected. Other cost savings which may be important for certain applications include those resulting from reductions in the post-chlorination dose, reductions in coagulant demand for colour or particle removal, reductions in sludge handling costs, and reductions in powdered carbon or oxidant dose to remove odour. The potential cost savings from these have not been well-documented, however.

2.4 Chlorine dioxide

Chlorine dioxide does not appear to have been successful as a replacement for chlorine. It does not form TTHM (see Fig. 3), it is an excellent bactericide and

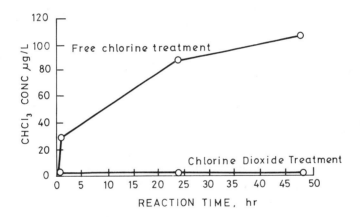

Fig. 3 — Chloroform formation in water containing 5 mg/l humic acid dosed with chlorine-free chlorine dioxide or free chlorine [10].

virucide and has been applied by many for oxidation of odour-causing compounds, but the major question that seems to be inhibiting its use concerns the health effects of chlorite and chlorate which form during its use. Application of doses sufficient to maintain a distribution residual in waters which have not been treated with activated carbon or a process to bring about biological stabilization will usually require a dose of chlorine dioxide that will cause the EPA-recommended minimum of 0.5 mg/l ($ClO_2 + ClO_2^- + ClO_3^-$) to be exceeded.

3. ADSORPTION

Adsorption is a common process in drinking water treatment plants. A 1977 survey [11] showed that about 25% of 600 plants (the 500 largest, plus 100 selected at random) in the United States used powdered activated carbon (PAC). PAC is added primarily for control of odour because it can be used only when odours are present, and the dose can be varied as the problem demands. At the time of the 1977 survey, there were only 45 known users of GAC and most of these were small plants ($<37\,800\,m^3$/day, or 10 MGD). The typical contactor was a rapid filter with all or part of the media replaced by GAC. The predominant reason for the use of GAC also was to control odour; when odour is present for a significant period of time during the year, GAC is often more cost-effective than PAC.

Since 1977 there have been some major findings which are serving as a force for more extensive use of adsorption. The problem of volatile organic chemicals (VOCs) has been discovered. There is widespread occurrence of groundwater supply contamination with compounds such as trichloroethylene and tetrachloroethylene, and unless an alternative source of water is available, the consumers are demanding removal of these compounds. There seems to be much greater concern with VOCs in groundwater than with organic contaminants in surface water simply because the public expects groundwater to be free of these industrial organics. Adsorption and aeration are two processes which can be used to remove VOCs and are being widely implemented for this purpose.

There also is increasing recognition of the vulnerability of many of our surface water supplies to organic contamination. In 1980, Swayne *et al.* [12] evaluated the water supplies for 1246 municipal utilities. Many of these supplies receive point discharge of wastewater above the point of water supply abstraction. As an indication of the situation, they reported that 20 cities with a total population of more than 7 million have surface water supplies containing 2.3–16% waste water during average flow conditions and 8–35% during low flow conditions. The distance from point of wastewater discharge to point of water supply abstraction was not given and, of course, will have an important effect on quality. However, it is interesting to note that few, if any, of these municipalities make any attempt to remove organic compounds other than those which cause odour or colouration. There are indications that some utilities are becoming aware of this problem and are beginning to consider the use of carbon adsorption processes in their drinking water plants for general removal of organic compounds. Public pressure may also be mounting because of the constant media discussion of hazardous waste and VOC contamination of water supplies.

The major problem confronting utility managers and regulatory officials who consider the possible use of adsorption for surface water treatment is quantifying the benefit that would be obtained for the expenditure. The majority of the organic compounds found in surface waters have not been identified and the long-term health effects of those that have been found are largely unknown. The frequency of occurrence of spills which might cause periodic high concentrations of harmful compounds is largely unknown. Thus, if adsorption is to be used, its health benefits will be largely unknown and design parameters and regeneration or replacement frequency will have to be set arbitrarily. A possible approach to solving this problem for GAC is given in Section 3.3

3.1 VOC removal: adsorption versus air stripping

Alternative processes for VOC removal include both adsorption and air stripping (aeration). Air stripping appears to be the method of choice in locations where no constraints are placed upon discharge of the gas stream into the atmosphere. Air stripping involves the passage of air through water thus allowing volatile contaminants to transfer from the liquid to the gas phase. Stripping efficiency depends upon the air volume applied per unit volume of water, the type of molecule as characterized by its Henry's constant, and the type of process, for example, e.g. spray tower, diffused aeration, or packed tower. Figure 4 presents a diagram that shows the

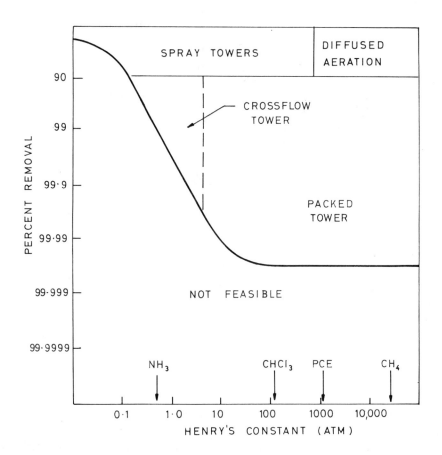

Fig. 4 — Diagram for the selection of a feasible aeration process for control of volatile compounds [13].

process which is likely to be best as a function of Henry's constant and removal efficiency [13]. If removal efficiencies less than 90% are required, diffused aeration can be used for the very volatile compounds and spray towers are applicable for the less volatile species. Removals greater than 90% will generally require a packed tower. The limits for the packed tower were determined by assuming that the height of a transfer unit is about 1 m and that 10 m represents the maximum economical height. As Henry's constant decreases, gas phase resistance becomes important and higher gas flows are required. The region boundaries should be taken only as approximate, and the diagram should be refined as more data become available. Most of the units installed to date are packed towers of the type shown in Fig. 5; in

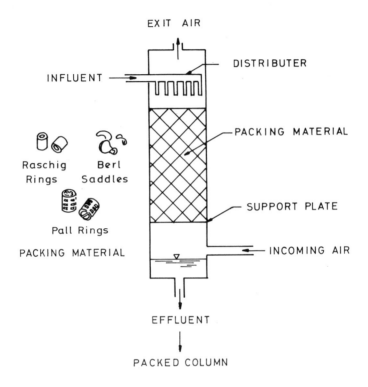

Fig. 5 — Aeration equipment [14].

these units the water is contacted with air as it flows in thin films over the packing media.

Aeration is not a costly process. Figure 6 shows costs (1981 $) as a function of plant size for 90% removal of trichloroethylene (see Hess in ref. 4). The removal of vinyl chloride and tetrachloroethylene is less expensive because they are more volatile, and carbon tetrachloride, 1,1,1-trichloroethane and 1,2-dichloroethane are more costly to remove because they are less volatile. Costs are shown for small plants because most supplies with a VOC problem are small.

Activated carbon also is highly effective for VOC removal (see Snoeyink in ref. 14 for a review). It is commonly used in single-stage contactors to treat small contaminated supplies. Prefabricated GAC pressure filters are available that can be

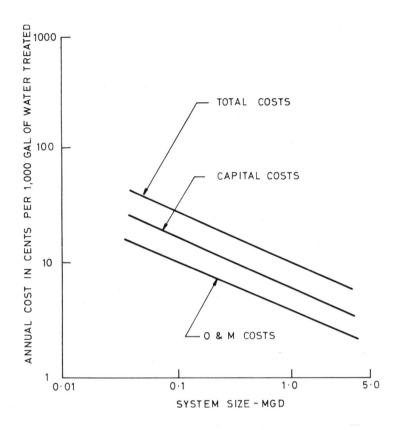

Fig. 6 — Costs for the packed column treatment technique [14]. (Note: system size represents average plant capacity.)

rapidly shipped and set up at sites in an emergency. It has the advantage over air stripping that non-volatile as well as volatile compounds can be removed, but while this may be important for supplies that are located near landfills, it has not been an important consideration for solving the general VOC problem encountered in the United States. GAC also has the advantage that the adsorbed compounds can ultimately be disposed of by incineration, whereas the gaseous effluent from air stripping is discharged to the atmosphere. The latter may not be permitted in some locations, thus necessitating the use of GAC alone, or an adsorbent to remove the VOCs from the air stripper's effluent gas. If this is required, it will add a substantial increment to the cost of air stripping.

GAC is more costly to use than air stripping if no gas stream clean-up is required for air stripping (see Snoeyink in ref. 10 for a review), Malcom Pirnie, Inc. calculated that the cost to reduce a 500 g/l concentration of trichloroethylene to 5 g/l was 6 ¢/m^3 (25 ¢/1000 gallon, 1981$), for a 7500 m^3/d (2 MGD) plant, 20 ¢/m^3 (82 ¢/1000 gallon) for a 750 m³/d (0.2 MGD) plant, and 40 ℓ/m³ (158 ℓ/1000 gallon) for a 75 m³/d (0.02 MGD) plant. Thus for a typical 1 MGD plant, the annual cost of GAC treatment is expected to be more than 2.5 times as much as air stripping.

The cost for addition of a GAC filter to clean up the air stream from an air stripping process is not available. Unknown factors that are required for determining the costs include sizing of the unit and regeneration requirements. Regeneration with steam may be possible because only the VOCs are present and these should be desorbed at a relatively low temperature.

3.2 GAC for surface water treatment

The primary application of GAC for surface water treatment to date is for odour removal for relatively small water supplies. The usual practice has been to replace all or part of the media in gravity, rapid sand filters with GAC, and then to use the GAC for the combined purpose of adsorption and particle removal. The required backwash frequency is similar to that used before GAC replacement. A bed life of the order of 1–3 years is often experienced, but the frequency and intensity of occurrence of odour during the period of use often is not well-documented, and there is evidence that if serious odours occur for long periods of time the carbon must be replaced or regenerated more frequently. Only two larger utilities, Manchester, NH, and Cincinnati, OH, have regeneration facilities and are operating their GAC filters to remove more organics than odour alone.

The application of GAC for adsorption only, following a rapid media filter, is an alternative that is commonly used in Western Europe. The required backwashing is less frequent, and thus there is less chance that the adsorption front will be upset leading to faster breakthrough. A disadvantage is that with less frequent backwash, higher organisms (such as rotifers and oligochaetes) may grow and be displaced into the product water (see Kruithof and van der Leer in ref. 15). More frequent backwashing should help to control such growths but this partially defeats the purpose of using a post-filter adsorber.

A procedure that could be used to control biological growth on GAC adsorbers is to use biological treatment as a process first in sequence in the treatment plant. As presented in Section 2.2, this process could be used to remove ammonia and biodegradable organics that would otherwise be biologically oxidized in the rapid filter and the GAC adsorber.

Preceding GAC treatment with ozonation produces good bactericidal and virucidal action, and oxidation of colour, odour, some inorganics and possibly other substances. It has the added effect of producing more biodegradable compounds in some waters; while this may be advantageous in providing better organic removal and extending the time between regeneration or replacement, it may be a disadvantage if the growths on the filters cause a problem. Excessive growth on filters should not be a problem if there is removal of bacterial substrate by biological treatment prior to ozone addition so that the amount of biodegradable matter entering the filters is small.

The cost of GAC treatment is a very important function of regeneration or replacement frequency. Cost data (1981 $) for a 37 800 m³/d (10 MGD) sand replacement system are given in Fig. 7, and the costs for a post-filter adsorber for the same size plant are presented in Fig. 8 [5]. As shown, the costs are reduced by a factor of

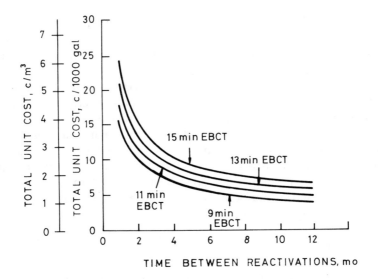

Fig. 7 — Total treatment unit costs vs. reactivation frequency for a 37 800 m³/day (10 MGD) GAC sand replacement system at various EBCTs.

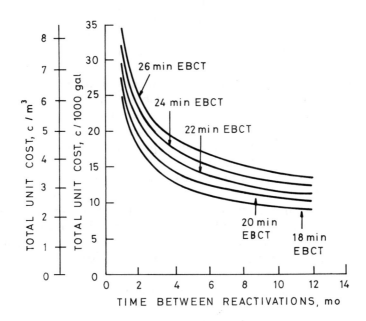

Fig. 8 — Total treatement unit costs vs. reactivation frequency for a 37 800 m³/day (10 MGD) GAC post-filter adsorber at various EBCTs.

two if the time between regenerations is increased from 2 to 8 months. Further, if addition of 2 mg/l of ozone before GAC increased the time between regeneration from 2 to 2.8 months, the decrease in GAC cost would equal the cost of adding the ozone.

A major problem facing designers and users of GAC technology is the establishment of treatment quality objectives that will allow the determination of when the GAC is exhausted. Without a quality objective, the regeneration frequency must be established arbitrarily, and development of optimum designs is not possible. Such an objective can be set for trichloromethane precursors, odour or colour, but we do not have a good basis for setting a standard for other trace organics. The principal concern about the latter groups of compounds is their long-term health effects; these are not known now and are not likely to be in the foreseeable future. Using GAC as a barrier may be a desirable way of dealing with this problem.

3.3 The barrier concept

We are familiar with the barrier concept in drinking water treatment because it is the basis by which we prevent contamination of drinking water with disease-causing organisms. We employ multiple barriers consisting of selection of the best possible source, coagulation-sedimentation, filtration, possibly pre-disinfection, post-disinfection and good maintenance of the distribution system. At each step microorganisms may be removed or prevented from entering the system, and the barriers are in place at all times, regardless of whether there are organisms present. However, no significant barrier to non-odorous trace organics exists in nearly all of our treatment plants which have supplies that are vulnerable to contamination. On the contrary, treatment practices such as chlorination introduce many unknown compounds into the water.

Employment of the barrier concept to control organic compounds would involve the use of GAC to treat supplies that are vulnerable to contamination. Process design and operation would be determined by a combination of cost and the degree of protection offered against selected compounds under a predetermined set of conditions. Perhaps three or four compounds, representative of classes of compounds that might at some time be found in the water supply, would be selected, and the adsorption capacity of GAC that had been in contact with the water for various times, for example, 3, 6, 9 months, would be determined for these compounds. The effectiveness of the barrier to the selected compounds could then be compared with the cost of operating the system with different regeneration frequencies. The benefit of the GAC could then be given in terms of its ability to adsorb spills. It is quite likely that GAC which has been in use for a year or more would still have good capacity for strongly adsorbing components such as some of the pesticides. Research would be required to determine whether this approach is feasible.

3.4 Resin-GAC and activated alumina-GAC

If the objective of treatment becomes one of removal of specific trace organics, rather than removal of TOC, for example, emphasis will be placed on the development of cost-effective systems that will accomplish that task. Strong base polystyrene resins with quaternary ammonium functional groups, and some weak acid resins, have a good capacity for natural organics. These resins can be regenerated with

NaOH solution. Natural organics will compete strongly with trace organics for adsorption sites on GAC so prior reduction of the concentration of natural organics may significantly increase the GAC bed life. An adsorption system consisting of a strong base resin to remove the natural organics, with *in situ* regeneration, followed by GAC to remove the trace organics (which the resin cannot remove) may be a cost-effective system, compared with GAC alone. Activated alumina has many of the same properties that strong base resins have for adsorbing natural organic matter and current research at the University of Illinois, Urbana, indicates that it can be regenerated with NaOH solution. Pre-ozonation also increases adsorption of natural organics on activated alumina. Similar processes may then be developed, with activated alumina followed by GAC to adsorb the trace organics.

Good cost and GAC bed life data are not available to thoroughly evaluate GAC versus resin-GAC or activated alumina-GAC. Some initial estimates of resin treatment costs are shown in Fig. 9 which, if valid would rule out the possibility of

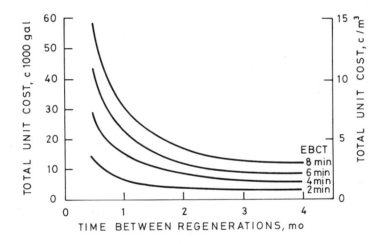

Fig. 9 — Total treatment unit costs vs. regeneration frequency for a 37 800 m³/day (10 MGD) anion exchange system at various EBCTs [5].

resin-GAC being more economical than GAC with a high regeneration frequency. A 37 800 m³/d (10 MGD) system operated with a contact time of 4 min and a two-week regeneration frequency would cost 7.2 ¢/m³ (29 ¢/1000 gallon). However, the majority of this cost is for resin replacement based on the assumption that 5% is lost with each regeneration. This loss is very high and I suspect additional research would show 1% or less to be a better estimate. More research is needed to determine regeneration frequency, loss on regeneration, cost of spent regenerant disposal, and the increased bed life of the GAC which follows, for both strong base resins and activated alumina before the process can be completely evaluated.

4. SUMMARY AND CONCLUSIONS

There is a major trend in the United States towards the sole use of monochloramine instead of free chlorine as a disinfectant. While this may be appropriate for many waters, it is inappropriate for some because of the reduced barrier to passage of virus into the distribution system and the lower effectiveness of monochloramine for preventing biological growths in some distribution systems.

The trend towards more extensive use of ozone is very desirable. Ozone is a good oxidant for colour and odour and a good bactericide and virucide, and a coagulation-filtration aid in some waters. It does not yield a lasting residual but it may be used in conjunction with chlorine, chlorine dioxide or monochloramine which can provide the needed residual. Care must be taken to control increased biological activity that may result from the application of ozone.

With the increased emphasis on reduction of disinfectant concentration as a means of controlling disinfectant by-product formation, more attention must be given to the production of a biologically stable water. One of the better alternatives for removing ammonia and biodegradable organics is to use a fixed-film biological process as one of the first processes in the treatment plant. Any organisms which are sloughed off can then be removed by the solids removal processes which follow. Other alternatives can also be used.

There is increased use of both air stripping and GAC adsorption for removing VOCs from contaminated groundwater. Air stripping is significantly cheaper than GAC adsorption so it is being used more frequently. However, if restrictions are placed on the discharge of contaminants to the atmosphere, carbon adsorption may come into more frequent use for controlling this problem.

GAC is rarely used to remove organics other than odorous compounds from surface water. Waste water is discharged to many sources of supply above the point of abstraction, however, and the fact that very few treatment plants that use these supplies have a barrier to passage of organic compounds is a serious problem. The data on the long-term health effects of low concentrations of organic compounds are minimal. Thus any treatment to lower the concentration will have to be done because of 'what might be'. Because of increasing public interest in the presence of hazardous wastes in the environment and in the quality, I anticipate that a significant number of communities will add GAC to their treatment plants in the next decade. Combinations of adsorbents, such as resin-GAC, or activated alumina-GAC merit more research because they may be cost-effective for this objective.

REFERENCES

[1] Davis, M. K. *et al.* The change of water treatment methods from chlorine to chloramines by water districts. *Contemporary Dialysis and Nephrology*, November 1984, pp. 24–26.

[2] Stevens, A. A. *et al.* Chlorination of organics in drinking water. *Journal American Water Works Association*, **68** (11), (1976) p. 615.

[3] National Academy of Sciences Safe Drinking Water Committee, *Safe Drinking Water and Health*, Vol. 2, National Academy Press, Washington, DC, 1980.

[4] Hack, D. J. State Regulation of Chloramination. *Journal American Water Works Association*, **77** (1) (1985) p. 46.

[5] Symons, J. M. *et al. Treatment Techniques for Controlling Trihalomethanes in Drinking Water*, U.S. Environmental Protection Agency Report, EPA-600/2-81-156, Cincinnati, OH, 1981.
[6] Rittmann, B. E. and Snoeyink, V. L. Achieving biologically stable drinking water. *Journal American Water Works Association*, **76** (10), (1984) p. 106.
[7] Short, C. S. *Removal of Ammonia from River Water-2*, Water Research Centre Technical Report TR3, Stevenage, Herts, 1975. Hodges, W. E. *et al.* Effect of Ozonation and Chlorination Pretreatment, American Society of Civil Engineers Conference, San Francisco, CA, 9–11 July, 1979.
[8] Fiessinger, F. *et al.* Advantages and disadvantages of chemical oxidation and disinfection by ozone and chlorine dioxide. *The Science of the Total Environment*, **18**, (1981), p. 245.
[9] Glaze, W. H. *et al. Evaluation of Biological Activated Carbon for Removal of Trihalomethane Precursors*, USEPA Report, Cincinnati, OH, 1982.
[10] Miltner, R. *The Effect of Chlorine Dioxide on Trihalomethanes in Drinking Water*, Thesis, University of Cincinnati, Cincinnati, OH, 1976.
[11] American Water Works Association Committee Report. Measurement and control of organic contaminants by utilities. *Journal American Water Works Association*, **69** (1977) p. 267.
[12] Swayne, M. D. *et al. Waste water in receiving waters at water supply abstraction points*. Report No EPA-600/2-80-044 United States Environmental Protection Agency, Cincinnati, OH, July 1980.
[13] Kavanaugh, M. C. and Trussell, R. R. *Air Stripping as a Treatment Process*, AWWA National Conference, 1981.
[14] KIWA-AWWA Research Foundation *Occurrence and Removal of Volatile Organic Compounds from Drinking Water*, Cooperative Research Report, American Water Works Association Research Foundation, Denver, CO, 1983.
[15] KIWA-AWWA Research Foundation *Activated Carbon in Drinking Water Technology*, Cooperative Research Report, American Water Works Association, Research Foundation, Denver, CO, 1983.

3.4

Water treatment research and development: a European viewpoint

T. H. Y. Tebbutt
School of Civil Engineering, University of Birmingham, PO Box 363, Birmingham, B15 2TT, UK

1. INTRODUCTION

Water is an essential constituent of the human environment and the presence of a reliable source is a vital factor in the establishment of a community. However, water is also capable of spreading disease and death as is only too obvious today in many areas of the developing world. The importance of an ample supply of safe water was recognized relatively early in man's development and the evidence of centuries old water supply systems can be seen in many countries. Archaeological investigations in Asia and China have exposed water distribution arrangements constructed several thousand years BC, and the Roman Empire developed comprehensive water supply installations for which some of the engineering structures still remain. Although the Roman engineers appreciated the value of bringing good quality water into cities, their main concern was to ensure a copious flow for the many fountains which were a feature of urban life at the time. Lead was a common material for water pipes and tanks, and it may be that the fall of the Empire was in part brought about by the lead content of the water supply. After the fall of the Roman Empire the countries of Europe entered a dark age lasting some 1500 years before public services such as water supply returned to the earlier standards. In the intervening period the inevitable consequence of the growth of urban communities around water sources with no effective provision for sanitation was an appalling death rate from water-related diseases. The Industrial Revolution, with its explosive growth of urban areas, forced the introduction of comprehensive water supply systems, followed fairly rapidly by the provision of sewerage systems and wastewater treatment facilities. Thus in many countries of Europe the majority of the population now has piped potable water supplies, although the percentage so served varies from about 60% in Austria to 99% in the UK. As in all parts of the world, rural communities are more likely to lack piped water supplies, so that differences in the percentage of the population served are largely due to different geographical distributions of populations. In Western Europe most, if not all, urban centres have ample supplies of safe potable water and significant outbreaks of water-related diseases are rare. This is, of course, very different from the situation in many areas of the world where large numbers of people suffer illness and death from water-related diseases such as cholera, typhoid, dysentery, schistosomiasis and other helminth infections. The fact that such diseases are almost unknown in Europe is one of the reasons why other aspects of water quality and treatment receive more attention. Obviously, climatic factors reduce the likelihood of occurrence of some of these diseases, but for others the potential danger always exists so that the primary aim in water supply is to attain satisfactory microbiological quality. However, the more recent concern at the presence of a whole range of micropollutants in drinking water has directed attention towards the removal of these contaminants. In parallel, the twin constraints of economic restrictions on capital expenditure and the rising costs of energy have brought considerable pressure on the water industry to become more efficient. Clearly the way in which efficiency is measured is a matter of some importance since consideration of purely economic factors could lead to a risk of diminished safety standards.

2. CONCEPTS IN WATER QUALITY

Because of its molecular structure, very high dielectric constant and low conductivity, water is capable of dissolving many substances so that it is an almost universal solvent. The corollary of this is that, strictly speaking, pure water does not occur anywhere naturally; rain water, surface waters and groundwaters all contain various amounts of both natural and man-made substances. Although rain water is normally considered to be relatively pure, the current interest in 'acid rain' has drawn attention to the sometimes undesirable characteristics of precipitation. As surface waters flow through the landscape they inevitably pick up substances from the catchment area. Infiltration of water into the ground often augmented by acid production due to soil microorganisms is likely to result in a solution of materials from the deposits through which the water passes. The discharge of domestic and industrial wastes to the environment will add to the natural level of impurities present in water. It is thus unrealistic to think of a 'pure' water, in chemical terms, as a practicable aim for potable supplies. Indeed, it is worth noting that pure demineralized water is not an acceptable drinking water. The natural ingredients in waters often serve a useful purpose in providing a balanced and palatable supply. It thus becomes necessary to adopt a rational approach to the specification of water quality by classifying water quality parameters into a number of types with particular characteristics.

(1) Toxic parameters that include a wide range of inorganic and organic substances capable of producing toxic effects in humans. For a particular substance the severity of the effect depends on the dose received, the period of exposure, and numerous other factors. Substances that are potentially toxic include: arsenic;

cyanide; lead; mercury; organochlorine and organophosphorus compounds; polycyclic aromatic hydrocarbons.
(2) Pathogenic microorganisms and microorganisms indicative of pollution. A large number of diseases are transmitted by the consumption of water containing the causative microorganism. Most of these diseases, such as cholera, typhoid and gastroenteritis, are faecal-oral in nature and occur as the result of faecal contamination of water supplies by sufferers from the diseases. The presence of normal non-pathogenic faecal microorganisms in water thus indicates a potential hazard. Viruses are pathogenic so that their presence in water poses a hazard although by their very nature the presence of viruses is difficult to demonstrate.
(3) Parameters undesirable in excessive amounts include a wide variety of substances some of which may be directly harmful in high concentrations; others may produce taste and odour problems. Some substances may not be directly troublesome, but are indicators of pollution. Constituents in this group include: fluoride; nitrate; iron; manganese; phenol; chloride; total organic carbon.
(4) Organoleptic parameters are those which are usually readily observable to the consumer but which normally have little health significance. Typical examples of organoleptic parameters are colour, taste, odour and turbidity.
(5) Natural physico-chemical parameters whose primary significance is that their presence in appropriate amounts ensures a balanced stable water although some may have limited health-related effects. Typical examples are: alkalinity; conductivity; dissolved oxygen; hardness; pH; total solids.

Classification of substances found in water in such a manner provides a basis for determining appropriate quality standards. Before considering the current European situation with respect to potable water, it is perhaps worth noting the pragmatic approach used in the UK until membership of the EC required more detailed legal requirements. The basic standard for water supply in the UK first established by the Waterworks Clauses Act in 1847, specified a supply of 'pure and wholesome water'. By the time of the Water Act of 1973 the definition had been abbreviated to a supply of 'wholesome water' possibly because of the term 'pure' was by then felt to be somewhat contentious. While it might be felt that such a vague definition of water quality was unacceptable in modern terms the reliance on wholesomeness did permit a sensible approach to quality control. Thus although there were no written standards in the UK a working understanding of wholesome water was taken to mean a supply which was free from: visible suspended matter; excessive colour; noticeable tastes and odours; objectionable dissolved matter; aggressive constituents; and, bacteria indicative of pollution. In reality, the essential requirements for a water supply are that it should be: potable i.e., safe to drink; palatable, i.e., pleasant to drink; and, suitable for all normal domestic and industrial uses.

It was perhaps inevitable that the increasing emphasis on environmental matters and the development of scientific means of assessing environmental risks has meant pressure to adopt quantitative standards for drinking water quality.

2.1 Water quality and health
There are two basic ways in which water quality can affect health; acute effects where the results of ingestion are almost immediate, and chronic effects where many years

of continued ingestion may be necessary before illness becomes apparent. Microbiological contaminants usually tend to produce acute effects, but most of the chemical and physical effects are of a chronic nature and are thus more difficult to detect in the early stages. With chemical and physical contaminants, acute doses are probably only likely to occur as the result of accidents in the catchment area and quite often the taste or appearance of the water would be such as to discourage consumption. Thus once the microbiological and, hopefully, viral quality of a water is satisfactory the main concern is likely to be with the long-term health hazards of chemical and physical constituents. It should, however, be noted that the World Health Organization (WHO) definition of health is 'a state of complete physical, mental and social well-being and not merely the absence of disease or infirmity'. Thus, in addition to substances for which an excessive intake would have an adverse effect on the consumer the above definition would also include the indirect health consequences arising from the presence of substances which make a water unattractive in appearance, taste or odour. This further complicates the position and often makes the establishment of quality criteria difficult to justify.

In relation to the microbiological quality of drinking water, the concept of indicator organisms, usually *E. coli* but sometimes *S. faecalis*, is well established, and the absence of such microorganisms in 100 ml of water is taken to imply that levels of pathogenic bacteria will be insignificant. There have been one or two suggestions that enteric pathogens have been found in water samples which did not give positive *E. coli* results, but the relative ease and reliability of coliform counts and the absence of any practicable alternative means that their use will continue. The potential hazard of viral particles in water is causing some concern but, although they can be found in many raw waters, cases of viruses being detected in treated waters in Europe are not well documented. The cynic might suggest that this may, in part, be due to the difficulty of undertaking viral examination of water samples.

When considering the establishment of standards for chronic exposure to chemical and physical contaminants in drinking water many problems arise. For many substances there is insufficient scientific information available to enable the rational establishment of water quality standards. Some substances may be essential for human nutrition but become hazardous at higher levels. The overall dietary intake must be considered and not just the intake in drinking water. The dose–response relationships of particular substances differ in that in some cases there is a threshold level below which no effect appears. Other substances, including probably all carcinogens, show no threshold effect so that any intake, no matter how small, is potentially hazardous. Materials behave in differing ways when taken into the body since adsorption in the gastrointestinal tract depends upon what other substances are present. The distribution of contaminants within the body may be widespread or concentrated in a single organ. Some materials may be rapidly excreted whilst others may be broken down into non-toxic end-products within the body. These factors, together with the variation in water intake due to age, climate and workload, serve to emphasize the problems to be found when trying to establish the level of a particular contaminant in water such that lifetime consumption of the water would not shorten the natural life span. Data from animal toxicity studies may provide some guidance but extrapolation of such data with large safety factors is again lacking in scientific justification. It is clearly important when determining the allowable levels of health-related constituents in drinking water to relate the possible health hazards to other health hazards in the community. There is, for example, little point in worrying about the possible carcinogenic properties of trace organics in water if the water is heavily populated by cholera bacteria. Wherever possible, therefore, some element of risk analysis must be incorporated into the process of determining water quality requirements.

2.2 Aesthetic aspects

Although, according to the WHO definition of health, even the aesthetic properties of drinking water can be considered as having health-related effects, it is usual to consider organoleptic parameters in a somewhat more relaxed manner. In the past the usual practice has been to reduce the level of such parameters as colour, taste and odour to the point where most consumers are not aware of any problem. It must, however, be appreciated that some people appear to be particularly sensitive to the appearance or taste of their drinking water so that it is probably impossible to satisfy the whole of the population. In these circumstances it must be accepted that complaints may be received from a small number of consumers and that it is economically and perhaps even practically not feasible to prevent these complaints. The growth in sales of bottled waters in the UK and other European countries is an indication of this factor, although, in fact, many purchasers may believe that the bottled, and expensive, water is 'better' than the mains supply. Unfortunately, this is not always true and the bacteriological quality of some bottled waters is not always satisfactory. With organoleptic parameters such as colour, taste and odour two particular problems arise; one related to the difficulties found with small supplies where removal of these parameters may not be feasible, the other to changes in the source of water used to supply an area which is becoming more common with large water authorities undertaking bulk transfers as part of an overall resource management strategy. It is normal for consumers to accept their usual water supply, with whatever organoleptic characteristics it may have, as 'good' water. Thus any change in characteristics observable to the consumer is likely to produce an adverse reaction.

2.3 Characteristics of importance to industry

Many industrial uses of water have less stringent requirements than potable supplies so that in general a drinking water will more than satisfy most industrial requirements. However, problems can arise with coloured supplies where industries such as soft drinks, high grade paper manufacture and the production of dyestuffs are involved. Changes in such parameters as hardness can be troublesome to industrial consumers who need to operate softening plants. Water abstracted for irrigation may have particular quality requirements in relation to trace concentrations of such materials as herbicides used for weed control in reservoirs and river channels.

2.4 Assessment of water quality

In line with the concept of 'wholesome' water the ideal in water quality assessment would be some form of 'black box' that would give an immediate indication of quality when presented with a sample. Developments in microelectronics occur at such a rate that it would be rash to predict that such a device will never be available but for the moment it must remain a gleam in the eye of a water scientist. The nature of the analytical problem can be gauged by the fact that over 600 organic compounds have been detected in drinking waters, most of them arising from various human activities. To this Pandora's box of contaminants must be added many inorganic compounds arising naturally and from human and industrial operations. The analysis of water samples for all the contaminants which might be present is clearly not possible in a routine monitoring programme. Many trace contaminants can only be detected by highly specialized analytical equipment and the actual analysis can take a considerable time to complete. It is thus necessary to rely for the regular assessment of water quality on relatively simple chemical and physical analyses such as alkalinity, ammonia nitrogen, colour, conductivity, hardness, odour, pH, phosphate, taste, turbidity, etc., possibly supplemented by a blanket parameter such as COD or TOC to indicate the organic contents. Depending upon the source of the water, routine tests for metals such as lead may be undertaken and it is becoming increasingly common to determine the aluminium content in treated water partly because of the indication it gives of the efficiency of the coagulation process but also because residual aluminium in a potable water supply can be a serious hazard to patients on kidney dialysis machines. When water is drawn from unpolluted and relatively protected sources, such as upland catchments and deep aquifers, the simple analyses mentioned above are generally perfectly satisfactory. Any major changes shown up by the analysis indicate the need for a more detailed examination to determine the reason for the change. With increasing use of lowland rivers and shallow aquifers for water supply purposes the level of natural and artificial contaminants will inevitably rise. Table 1 illustrates this point with data from the

Table 1 – Typical water quality in the River Severn

Distance from source (km)	Total dissolved solids (mg/l)	Conductivity (μS/cm)	Alkalinity (mg/l)	Chloride (mg/l)	Total carbon (mg/l)	Ammonia N (mg/l)	Nitrate N (mg/l)
16	74	138	16	8	6	0.02	0.6
64	166	233	71	24	15	0.04	0.9
140	307	437	120	48	20	0.04	2.3
175	361	479	132	56	22	0.10	2.7
200	417	562	138	60	23	0.52	2.9

River Severn, UK. Such sources are much more likely to become contaminated as the result of accidents in the catchment area. Pollution of water courses by motor vehicle accidents, industrial mishaps and careless use of agricultural chemicals are relatively common events which cause considerable concern to water supply authorities with abstractions below the point of pollution discharge. In a well-regulated society, such accidents should always be reported to the appropriate organizations so that suitable precautionary action can be taken. Such an early warning cannot always be relied upon however; 'accidents' happen at night and even if warnings are received the exact nature of the contaminant and the time for it to reach intakes may not be immediately known. There is increasing interest in biological monitors to provide advance warning of potentially toxic doses of pollutants in water supplies. These can involve fish kept in instrumented aquaria which detect changes in the respiration rate or behaviour of the fish. Another type employs an ammonia electrode to monitor the output from a small bed of nitrifying bacteria. These bacteria are sensitive to many environmental factors and a decrease in their activity will mean that not all the ammonia nitrogen in the raw water will be converted to nitrate nitrogen. Thus residual ammonia nitrogen in the raw water will be detected by the ammonia electrode, the signal from which can be used to actuate an alarm system. It is, of course, then necessary to subject the water to a detailed analysis to determine the contaminant responsible for the alarm and to ascertain its significance in relation to potable water supply.

In the UK and a number of other European countries it is common practice to provide several days of bank-side storage when abstracting from a river likely to be subject to accidental pollution. Such storage enables the intakes to be closed to allow a plug of contaminated water to flow by, or if not detected, the contaminant is diluted by the contents of the storage reservoir.

2.5 Water quality standards and guidelines

Although specific water quality standards did not exist in the UK until relatively recently they have been used in several parts of the world for a considerable time. The concept of international standards was introduced by the WHO in the early 1950s and resulted in the publication of International, and later, European Drinking Water Standards. International standards were considered to be the targets for all supplies whereas the European standards were in some aspects more stringent, partly because of the increased potential for environmental pollution by chemicals in the industrialized countries of Europe and partly because it was felt that wealthy countries could afford to adopt more stringent standards. The WHO standards were, of course, only advisory in nature since they had no legal basis in individual countries but they were often adopted as the basis for water quality requirements. The latest WHO recommendations incorporate Guidelines for Drinking Water Quality. These are based on the use of Action Levels which have been formulated with a view to safeguarding public health from potentially toxic substances for a lifetime. If the concentration exceeds the Action Level the reason must be investigated and appropriate remedial action taken.

In the European Community the situation has changed in that there are a number of Community directives that are concerned with water quality. These will have legal force in Community countries and deal with raw water for potable supplies, potable water itself, bathing water, freshwater fisheries and sea water in shellfish areas. In most cases the directives set out two values; guide (G) values which are targets and mandatory (I) values which are maximum allowable concentrations (MAC). Tables 2 and 3 give examples of the EC water quality directives. It should be noted that in the case of raw water quality the directives relate the values to the type of treatment to be employed. With conventional water treatment processes it is necessary to classify water contaminants into three general types.

(1) Contaminants that cannot reliably be removed. These include; arsenic, barium,

Table 2 — Some examples from EEC Directive 75/440. Characteristics of surface water intended for potable water abstraction

	Treatment type[a]					
	A1		A2		A3	
Parameter (mg/l except where noted)	Guide limit	Mandatory limit	Guide limit	Mandatory limit	Guide limit	Mandatory limit
pH units	6.5–8.5		5.5–9.0		5.5–9.0	
Colour units	10	20	50	100	50	200
Suspended solids	25					
Temperature (°C)	22	25	22	25	22	25
Conductivity (μS/cm)	1000		1000		1000	
Odour (TON)	3		10		20	
Nitrate (as NO_3)	25	50		50		50
Fluoride	0.7–1.0	1.5	0.7–1.7		0.7–1.7	
Iron (soluble)	0.1	0.3	1.0	2	1.0	
Manganese	0.05		0.1		1.0	
Copper	0.02	0.05	0.05		1.0	
Zinc	0.5	3.0	1.0	5.0	1.0	5.0
Boron	1.0		1.0		1.0	
Arsenic	0.01	0.05		0.05	0.05	0.1
Cadmium	0.001	0.005	0.001	0.005	0.001	0.005
Chromium		0.05		0.05		0.05
Lead		0.05		0.05		0.05
Selenium		0.01		0.01		0.01
Mercury	0.0005	0.001	0.0005	0.001	0.0005	0.001
Barium		0.1		1.0		1.0
Cyanide		0.05		0.05		0.05
Sulphate	150	250	150	250	150	250
Chloride	200		200		200	
MBAS	0.2		0.2		0.5	
Phosphate (as P_2O_5)	0.4		0.7		0.7	
Phenol		0.001	0.001	0.005	0.01	0.1
Hydrocarbons (ether soluble)		0.05		0.2	0.5	1.0
PAH		0.0002		0.0002		0.001
Pesticides		0.001		0.0025		0.005
COD					30	
BOD (with ATU)	<3		<5		<7	
DO percent saturation	>70		>50		>30	
Nitrogen (kjeldahl)	1		2		3	
Ammonia (as NH_4)	0.05		1	1.5	2	4
Total Coliforms/100 ml	50		5000		50 000	
Faecal Coliforms/100 ml	20		2000		20 000	
Faecal Streptococci/100 ml	20		1000		10 000	
Salmonella		absent in 5 litre		absent in 1 litre		

[a]Treatment types.
A1, simple physical treatment and disinfection; A2, normal full physical and chemical treatment with disinfection; A3, intensive physical and chemical treatment with disinfection.
Mandatory levels, 95% compliance; 5% not complying should not exceed 150% of level.

cadmium, chloride, chromium, copper, cyanide, fluoride, hydrocarbons, lead, nitrate, phenol, sulphate, zinc.
(2) Impurities that can be removed reliably within certain limits. These include;

Table 3 — Some examples from EEC Directive 80/778. Quality of water intended for human consumption

Parameter	Expressed as	Guide level	Maximum admissible concentration
Colour	mg/l Pt/CO scale	1	20
Turbidity	mg/l S$_i$O$_2$ scale	1	10
Odour	Dilution number	0	2(12°C)3(25°C)
Temperature	°C	12	25
pH	unit	6.5 pH 8.5	
Conductivity	µS/cm (20°C)	400	
Chloride	mg/l as Cl	25	
Sulphate	mg/l as SO$_4$	25	250
Calcium	mg/l as Ca	100	
Magnesium	mg/l as Mg	30	50
Sodium	mg/l as Na	20	175
Total dissolved solids	mg/l		1500
Nitrate	mg/l as NO$_3$	25	50
Ammonia	mg/l as NH$_4$	0.05	0.5
Phenols	µg/l as C$_6$H$_5$OH		0.5
Boron	µg/l as B	1000	
Iron	µg/l as Fe	50	200
Manganese	µg/l as Mn	20	50
Phosphorus	µg/l as P$_2$O$_5$	400	5000
Fluoride	µg/l as F		1500(12°C)700(25°C)
Arsenic	µg/l as As		50
Cadmium	µg/l as Cd		5
Cyanide	µg/l as CN		50
Mercury	µg/l as Hg		1
Lead	µg/l as Pb		50
Pesticides (total)	µg/l		0.5
PAH	µg/l		0.2
Total coliforms	MPN/100 ml		<1
Faecal coliforms	MPN/100 ml		<1
Total colonies (37°C)	per ml	10	
Total colonies (22°C)	per ml	100	

inorganic and organic suspended solids, some soluble organics, calcium, carbon dioxide, iron, magnesium, manganese, microorganisms.
(3) Impurities that may interfere with water treatment processes. These include; ammonia, phosphate, synthetic detergents.

With tne first class of contaminants the MAC in the raw water must be the same as that acceptable in potable water. It may, however, be possible to use raw water with higher than permissible levels of contaminants if blending with a better quality supply can be practised. With the other two types of contaminant the maximum levels allowable in both raw and treated waters will be influenced by the performance of the treatment plant and the economics of the operation.

The requirements of the Water for Human Consumption directive are causing some concern since there are a number of values which appear to be almost irrelevant to the protection of consumers. An example is the case of colour with a G value of 1°H and an I value of 20°H. It has been estimated by a large UK water authority that full compliance with the directive could cost about £30m for a population of 7 million. Much of this expenditure would produce only an improvement in the appearance of the water and marginal reductions in some naturally occurring inorganic substances. The logic of undertaking expenditure with such questionable returns is difficult to establish when compared with the well-established hazards to life of motor vehicles and cigarette smoking. The credibility of mandatory standards is not helped when it would appear that some levels are based on the minimum detectable concentration of a particular contaminant, or, in the case of the freshwater fishery directive, set out as unacceptable, concentrations of substances which are commonly found in healthy fishing waters. It is surely important that whatever water quality standards are adopted they are seen to be based on logic and are possible to attain. If this is not done the standards will almost certainly be ignored, whereas a more realistic level would be accepted.

3. DEVELOPMENTS IN WATER TREATMENT

The water supply industry is traditionally conservative so that changes in treatment systems tend to be evolutionary rather than revolutionary. The environmental and economic pressures that have an increasing influence on the way in which the industry operates do, however, mean that traditional solutions are likely to be more frequently questioned. There have thus been a number of significant developments in several areas of water treatment and these are discussed briefly below.

3.1 Coagulation and clarification

Most river-derived abstraction schemes employ chemical coagulation for the removal of turbidity, and the adsorptive properties of hydroxide flocs are also often used for colour removal from upland catchment sources. Problems associated with poor floc settling characteristics have been largely obviated by the availability of polyelectrolyte coagulant aids. These are large organic molecules in polymerized forms that act as floc binding and strengthening agents when used in small doses, usually <1 mg/l, in addition to primary coagulants such as aluminium sulphate. Some of these polyelectrolytes are natural compounds, e.g., starch derivatives, but many are synthetic substances like polyacrylamides and carboxymethyl celluloses. In the UK the possible health hazards of such organic additives are given careful consideration by a government committee, which has approved some 150 products for use in potable water treatment subject to specified maximum dose levels. The availability of these coagulant aids has resulted in considerable benefits in terms of improved settled water quality, better filtrate quality and, not infrequently, lower treatment costs.

The simplicity of vertical-flow sludge blanket settling tanks continues to encourage their use in many plants, although the earlier inverted pyramid units have to some extent been supplanted by flat-bottomed tanks which are cheaper to construct and appear to offer similar performance to that of the pyramidal tanks. The French vacuum-pulsed flocculation system for upward flow tanks is becoming widely used, and in side-by-side comparisons with circular combined flocculation-sedimentation tanks has shown a better settled water quality. Although the adoption of polyelectrolytes has usually led to improved performance of settling basins there are some cases where the floc settling characteristics are still poor. There has thus been a growth in the number of flotation units commissioned for solids/liquid separation in water treatment plants. Most of these employ the dissolved air flotation system and recycle *ca.* 10% of the treated water. The process becomes attractive when floc settling velocities are around 2 m/h or less, and since the requirement is for relatively poor settling properties, coagulant aids are not usually necessary or indeed desirable. With all coagulation operations optimum performance depends upon close control of chemical doses. The traditional jar-test procedure still plays a major role in determining the appropriate coagulant dose but there have always been difficulties in quantitatively transferring the laboratory results to the full scale situation. Some progress has been made in producing site-specific algorithms that predict the optimum coagulant dose on the basis of easily measured parameters in the raw water such as alkalinity, colour and turbidity. A promising development in the control of coagulation has been made at University College, London, where an on-line continuous flocculation monitor has been developed for commercial production. This device allows real-time observation of changes in chemical dosage and their effects on floc size and properties.

For softening operations upward flow pellet-type reactors have become popular. These units use hard calcium carbonate pellets suspended in a fluidized bed to catalyse further precipitation with the great advantage of producing a dense, rapidly draining sludge with significant agricultural potential. However, the recent evidence of a relationship between soft water and certain forms of cardiovascular disease has suggested that for domestic supplies softening below 150 mg/l (as $CaCO_3$) is probably not desirable.

3.2 Filtration

Filtration through deep porous-media beds provides the final polishing treatment for coagulated waters and the main treatment for low-turbidity raw waters. It is a highly complex process involving physical, chemical and sometimes biological actions. Because of this complexity it is not easy to establish optimum design and operating parameters, although a considerable research effort in several countries has done much to highlight the important factors in filter performance. The concept of improved usage of the voids in a filter bed has encouraged attempts to design units so that the head loss and filtrate quality limits are reached at the same time. Single-medium beds now tend to use somewhat larger sand sizes than in the past and many filters use dual anthracite-sand beds, either as original equipment or as a way of up-rating the capacity of existing units. Upward and radial flow filters offer theoretical advantages in the form of more effective use of the voids, but are inherently more unstable than the conventional down-flow mode so that their use as the sole filtration stage for the production of potable water is not common.

As an alternative to chemical coagulation followed by rapid filtration, a number of major installations use direct application to rapid filters followed by a secondary slow-filtration stage. It is interesting to note that the apparently obsolete slow sand filtration process has witnessed a resurgence of interest in some areas as a result of its ability to provide a high quality water with in-built oxidation of at least some dissolved organics because of biological activity, an important feature of the slow filter.

The backwashing of rapid filters has largely standardized on the use of air scour prior to, or concurrently with, the water wash with bed expansions of 10–15%, higher expansions being considered to decrease the cleaning action as well as being more costly and tending to result in losses of sand.

3.3 Disinfection

In view of the potential hazards arising from the presence of microorganisms in water supplies, disinfection must be considered as the essential treatment process. The great majority of potable water supplies in Europe use chlorine as the disinfectant and residuals are normally kept to a level which is not detectable by the consumer. This contrasts with some other parts of the world where it is thought that water cannot be properly treated unless it has a pronounced taste and odour of chlorine. Concern about trihalomethanes and other organochlorine compounds present in water supplies, originated in the United States but is also found in Europe. However, most water suppliers in Europe do not believe that there is a major problem with these substances. This is partly because of the more restricted use of chlorine in Europe. Disinfection of sewage effluents before discharge to watercourses is generally not favoured because of the potential by-product formation and the effects on aquatic life. Pre-chlorination of raw waters is used as sparingly as possible, with the main aim of controlling biological growth in the pipelines and treatment units only and not to provide oxidation of organic compounds. Chlorine dioxide is in use in some plants, usually with waters having taste or odour problems, since the cost means that it is economically unattractive if used solely for disinfection purposes.

Ozone has had a long history in Europe and is the second most popular disinfectant. More efficient generators have reduced the cost disadvantage of ozone in relation to chlorine, and the colour bleaching properties of ozone have led to its adoption for a number of upland supplies where the alternative method of colour removal would have been chemical coagulation. There is, however, some evidence that there is a partial return of colour after ozonation and the lack of persistence of ozone can cause problems in the distribution system unless a small chlorine dose is added before the water leaves the treatment plant. Ozone forms ozonides with organic impurities but as yet little is known about the properties of these by-products. The reactions of chlorine with water and its impurities have been the subject of an enormous amount of observation and research. It would therefore be unscientific to replace chlorine as the first choice for disinfecting water until the reactions of any alternative disinfectant have received at least as much attention as those of chlorine and its compounds.

For small water supplies, disinfection often causes problems because of the difficulty of providing reliable fail-safe chemical feeders. The EEC requirement for low colour levels in potable water has resulted in the adoption of simple UV lamps to provide both a degree of colour removal and the required disinfection for small rural water supplies. Such units require minimum levels of maintenance and cannot overdose the water.

3.4 Removal of micropollutants

Although the need to remove micropollutants because of potential health hazards is open to argument, there is no doubt that troublesome tastes and odours, often due to micropollutants in potable waters are by far the most common cause of complaint. Micropollutants may be naturally occurring compounds resulting from algal metabolism or from the activities of bacteria and fungi or they may arise in the form of residues from agricultural and industrial operations. Most lowland surface waters and increasing numbers of shallow groundwaters will reveal the presence of a range of organic contaminants if subjected to sophisticated analysis by gas chromatography–mass spectrometry techniques. Reactions between these micropollutants and chemicals used in the treatment process, particularly disinfectants, can produce different forms of micropollutants which may have increased health hazard potential and often show amplified taste and odour characteristics.

It is generally believed in Europe that the best way of dealing with micropollutants is to prevent their entry to the water source if at all possible. When micropollutants are already present in raw water, as much as possible of the contamination is removed before disinfectants are added to water. This ensures that organochlorines or other by-products are kept to a minimum. In the Netherlands polluted raw waters are subjected to a complex treatment chain involving rapid and slow filtration, aeration and artificial recharge into sand dunes to remove organic compounds and microorganisms so that only a small final disinfection dose is required to produce a safe supply. Activated carbon is usually considered the ultimate solution for taste and odours due to micropollutants. For occasional problems powdered carbon added to the coagulation or filtration stages is common. With more continuous problems, granular activated carbon beds following sand filters can be employed but the costs of carbon and its regeneration do not encourage its use unless absolutely necessary. In any event it must be emphasized that activated carbon does not provide a universal solution for the removal of trace organics since the adsorbability of various organic substances differs widely.

Other methods used for taste and odour removal include aeration, which can be quite effective for volatile contaminants, and various forms of oxidation using chlorine or ozone. Whilst these latter techniques may succeed in removing the offending tastes and odours they can, of course, produce residual organic substances which may be undesirable because of possible long-term health hazards. As with many areas of environmental contamination it is therefore usually preferable to prevent the undesirable material entering the raw water rather than to try to remove it once it is there.

Although not strictly speaking a micropollutant, nitrate nitrogen is receiving a considerable amount of attention in European water supplies. Intensive agricultural activity and sewage effluent discharges contribute significant amounts of nitrate nitrogen to lowland rivers and lakes and also possibly to groundwaters. The presence of nitrate nitrogen in bottle feeds for young babies can result in methaemoglobinaemia which can be fatal if untreated. For rural areas it may be possible to provide the babies at risk with nitrate-free bottled water; for larger communities this would not be satisfactory. Water containing more than 11.3 mg/l of nitrate nitrogen must thus either be blended with a lower nitrate nitrogen content supply if available or alternatively the nitrate must be removed. This can be achieved by biological denitrification with methanol as a carbon source, which may itself have a potential organics residual problem unless accurately controlled, or by ion exchange which tends to be non-selective, so that the economics of the process are influenced by the anion content of the water.

3.5 Sludge treatment and disposal

Until relatively recently the residuals arising from water treatment operations, i.e., clarification sludges and filter backwash flows, were given little attention at the design stage and in operational terms were often dealt with on an *ad hoc* basis. It was by no means unknown for sludges and washwater to be discharged to water courses below the abstraction point, on the dubious argument that the material had come out of the water in the first place. In England and Wales the formation of multi-purpose water authorities resulted in a more responsible attitude to the handling and disposal of water treatment residuals. One arm of an authority could hardly be seen to be polluting water sources the protection of which was the responsibility of another arm of the same body.

The major problem in water treatment sludges is their low solids content, so that there is a primary requirement for effective dewatering processes. Gravity thickening followed by pressure filtration to produce a final cake of 25–40% solids is now generally recognized as an appropriate system. Conventional hydroxide sludges, arising from chemical coagulation operations to remove turbidity and colour, have no beneficial use so that disposal to tip of the dewatered material is the most likely solution. Alum recovery is feasible by means of acid treatment but the cost of the recovery process and possible operational problems owing to a build-up of intractable colour in the recycled chemical make the technique unattractive. Direct spray application of liquid coagulation sludges to waste land is environmentally unattractive. In the case of sludges from softening operations there can be a potential market in the form of agricultural use of calcium and magnesium sludges. This is, however, a minor area of sludge production and in view of the evidence that softened waters have some influence on the incidence of certain forms of cardiovascular disease the use of softening for domestic supplies may well decline.

3.6 Operation and control

Water treatment plants can of course be quite complex systems involving considerable capital and operational expenditure. There is thus a need for effective design and operational strategies based on appropriate applications of system analysis. Energy consumption is receiving a great deal of attention since power costs are often a major part of operating expenditure. On-line control of chemical dosage, sludge handling, filter backwashing and other activities are all being used to increase efficiency. The adoption of expert systems and intelligent knowledge-based systems is in an early stage in the water industry but clearly has potential value for both design and operation decision-making. In most water distribution systems there are significant losses of water, often of the order of 25%, via leakage and waste. A great deal of work is now being implemented to reduce this loss by such measures as automatic pressure regulators, which ensure that high pressures do not develop during periods of low flow. Leakage control measures can often reduce the need for increased treatment capacity as well as saving operating expenditure. Developments in microelectronics are so rapid that predictions are likely to be overtaken by events very rapidly. The concept of demand management by variable tariffs, as practised by other utilities such as electricity supply, could well play an important role in the future management of water treatment and supply systems.

4. CONCLUSION

In what is inevitably a personal view of the environmental technology aspects of water supply and treatment it would be too much to expect complete coverage of the subject area. The aim of this chapter has been to set out what appear to the author to be some of the most significant features of the European water treatment scene. In relation to future developments and research needs the following subjects appear to require particular attention:

(1) Realistic assessment of the risks associated with the presence of trace contaminants in water supplies;
(2) A detailed study of the behaviour of alternative disinfectants in water with particular reference to the characteristics of any by-products as a result of disinfection;
(3) Effective methods for the control of organoleptic parameters such as colour, taste and odour;
(4) Reduction of water demands by improved methods of leakage control and demand regulations;
(5) Applications of information technology and expert systems to provide more effective design and operational procedures.

Research and development for the water industry can be undertaken by a number of types of organization. Because of the many factors that influence water quality, research in this area requires a multidisciplinary approach drawing together, as appropriate, the expertise of civil engineers, biologists, chemists, physicists, electronics and computing specialists. This could well indicate that a university environment would be appropriate for much of the fundamental research needed in the water industry. Certainly many European educational institutions play a significant role in water research with funding provided by central government bodies or in the form of contracts from organizations in the water industry. Increasing emphasis is placed on collaboration between the educational and industrial sectors, and when assessing research grant proposals the presence of a significant industrial contribution in terms of cash, manpower and materials is an important aspect. The university approach to research is often suitable for the initial study of speculative proposals, or for what may be termed defensive research. Most European countries have government or public authority research establishments for water research. These establishments usually have a dual role, undertaking strategic research to provide a better understanding of topics deemed to be of major general importance to the water industry, and providing a custom-made research service to examine specific problems on an *ad hoc* basis as and when they occur. These latter investigations are often carried out in collaboration with staff from the organization which has brought the problem to light. Although it may sometimes be possible to involve university staff in these *ad hoc* studies it is more likely that such collaboration will arise within a programme of strategic research. This permits the academic researcher to carry out fundamental work which is, however, related to the needs of industry. Thus a better understanding of the manner in which treatment processes operate gained from carefully controlled laboratory studies can result in more efficient full-size units. The increasing need for greater performance efficiency and lower energy consumption in water treatment systems coupled with demands for more stringent assessment of environmental factors can only be satisfied if a strong research base provides the information for logical decision making.

BIBLIOGRAPHY

Council of the European Communities. Directive of 16 June 1975, concerning the quality required of surface water intended for the abstraction of drinking water in the Member States. Directive 75/440/EEC. *Official Journal*, 25 July 1975.

Council of the European Communities. Directive of 15 July 1980, relating to the quality of water intended for human consumption. Directive 80/778/EEC. *Official Journal*, 30 August 1980.

Institution of Water Engineers and Scientists. *The Water Treatment Scene — The Next Decade*, IWES, London, 1979.

Kenny, A.W. The effect of EEC directives on water supplies. *J. Inst. Wat. Engrs. Scits.*, **34** (1980), 42.

Tebbutt, T. H. Y. (ed.). *Water and Waste Research — The Way Ahead*, Science and Engineering Research Council, Swindon, 1982.

Tebbutt, T. H. Y. *Relationship Between Natural Water Quality and Health*, Technical Documents in Hydrology, Unesco, Paris, 1983.

World Health Organization. *Guidelines for Drinking Water Quality*, WHO, Geneva, 1984.

Part 4
Wastewater treatment

4.1

Research needs for wastewater treatment and management of resulting residues

R. I. Dick
School of Civil and Environmental Engineering, Cornell University, Hollister Hall, Ithaca, New York 14853, USA

1. INTRODUCTION

1.1 Background

The need for new knowledge for controlling water pollution from municipal and industrial wastewater discharges has never been greater than now. There is a substantial backlog of unresolved water pollution control problems, and, in recent times, new problems have developed more rapidly than old ones have been solved.

Such has been the history of water pollution control. Problems arise, they assume crisis proportions, available solutions are implemented, and then decades are spent in seeking fundamental understanding of the problems and refining or revising prior hasty solutions. Thus, the research needs suggested in this chapter address a combination of old and new problems.

Few fields of scientific or engineering enquiry match the complexity and diversity of contemporary water pollution control. Difficult and interrelated chemical, physical and biological problems were always involved in water pollution control, but as industry explored its capabilities, problems of controlling water pollution from resulting products and dregs complicated an already complex problem.

Changes other than industrial development have also complicated water pollution control technology. Increased public expectations of environmental quality have caused previously acceptable solutions for water pollution control to become unacceptable. And previously accepted materials (like carbon tetrachloride and polychlorinated biphenyls) have come to be known as human health threats.

1.2 Areas of research

This review of pollution control research needs includes consideration of research for wastewater treatment processes and for treatment, utilization and disposal of the residues produced in the course of wastewater treatment. Because of the wide diversity in industrial wastewater treatment practices, this general review of research needs is focused on typical municipal wastewater treatment practices.

Traditionally, environmental engineering education, research and design have been oriented to the removal of contaminants from wastewaters. Experience, however, has shown that the management of the sludges produced by wastewater treatment costs approximately as much as the wastewater treatment processes that generate the sludge. Sludge management processes are of comparable complexity and diversity as processes for wastewater treatment, and, furthermore, less is known about them because of historical neglect. Hence, readers will find sludge management commands about as much attention in this chapter as does wastewater management.

1.3 Problems not embraced

Important wastewater treatment and residue management informational needs are not considered in this chapter. Even the broad swathe of wastewater treatment technology does not encompass significant problem areas. Among the many related areas not embraced by this chapter are source control and pretreatment of industrial wastewaters, design and operation of sewerage systems, laboratory methodology, toxicology, epidemiology, hazardous waste control, nonpoint source control, fate and effects of treated effluents on receiving waters and lands, river basin management, land use management, environmental impact assessment, water quality planning, water quality criteria, water quality regulations, risk assessment, and pollution control financing. Clearly the multitudinous research requirements identified here must be expanded many-fold to embrace the entire area of water pollution control.

2. WASTEWATER TREATMENT

Large expenditures will continue to be made to control discharge of traditional municipal and industrial pollutants like biochemical oxygen demand and suspended solids; hence research to improve performance and cost efficiency are justified. More newly recognized threats presented by microcontaminants justify extensive research concerning their control in conventional and nonconventional treatment processes.

Recent surveys of developments and research needs in wastewater treatment in the United States include the proceedings of a conference, *Fundamental Research Needs for Water and Wastewater Systems* [1], a Water Pollution Control Federation Committee report, *Research Needs Associated with Toxic Substances in Wastewater Treatment Systems* [2], and a series of Research Planning Task Group Studies sponsored by the University of Illinois Advanced Environmental Control Technology Research Center [3,4,5,6].

2.1 Traditional contaminants

While conventional wastewater techniques have benefitted from as much as a century of development, it is easy to identify significant limitations in understanding process performance. The common use and expense of these processes justifies far more research than has occurred in recent years in the United States.

2.1.1 Suspended solids removal

The economics of treating large volumes of wastewater favour chemical or biological conversion of contaminants into suspended form for removal by solids–liquid separation processes. Thus, the oldest of all wastewater treatment processes, suspended solids separation, probably will continue to be the most commonly used method of wastewater treatment. Readers are referred to an assessment of solids–liquid separation developed by the Engineering Foundation [7] for a review of the current state of the art and research needs in this field.

If the most prevalent means of separating contaminants from wastewater is to be solids–liquid separation, then it is prudent to learn much more about the fundamental nature of the particles to be separated. This applies to particles contained in raw wastewaters as well as those formed in chemical and biological treatment processes. Means for regulating the growth of agglomerate particles to assure development of dense, tough agglomerates would aid performance of wastewater treatment and subsequent sludge management processes. This requires improved understanding of particle agglomeration and floc breakup mechanisms. Methods for measuring agglomerate properties such as size, density, and strength as they exist in treatment plants, are necessary to advance current understanding of particle agglomeration, agglomerate properties, and breakup of agglomerates in shear fields.

In spite of the long and widespread use of sedimentation basins in wastewater treatment, analysis of their performance reveals substantial deviation from expectations based on theory [8]. With a given distribution of particle sedimentation velocities, sedimentation tank efficiency should diminish with increased hydraulic loading per unit of surface area. In fact, other factors affect performance more significantly. This failure of settling tanks to conform to rational expectations is probably related to factors such as density currents, inlet energy dissipation, scour, wind effects, and flow distribution. Much more must be learned about the control of these and similar effects in order to make settling tank performance approach that of 'ideal' settling basins [9] upon which traditional design concepts are based.

2.1.2 Removal of biodegradable organic compounds

While the removal of biodegradable organic materials has long been accomplished effectively at large scale in municipal and industrial wastewater treatment plants, much additional research is justified by the continuing common need for such treatment, by the high cost of treatment, and by the lack of current understanding of important mechanisms controlling process performance. In this author's opinion, elaboration of kinetic formulations for removal of biodegradable organic compounds, which has fascinated so many researchers in the past quarter of a century, deserves only low priority in the future. While such work has improved understanding of mechanisms of biological wastewater treatment, it has not addressed major problems in biological wastewater treatment.

Conventional activated sludge treatment ordinarily accounts for the most use of energy in municipal wastewater treatment, and means for reducing this energy consumption and cost are needed. Research addressing this problem will include consideration of alternative reactor configurations as well as means for monitoring and controlling conventional reactors to reduce energy consumption.

Bioflocculation and solids separation are major factors limiting current ability to control the performance of suspended growth biological wastewater treatment processes. It is necessary to learn more about the conditions that allow specific types of organisms to thrive in biological reactors, about population dynamics, about extracellular polymer production, and about agglomerate particle development and disruption (see previous section) so as to design and operate suspended growth biological processes that consistently flocculate and settle well. A preliminary indicator of the direction such research may take is the use of 'selectors' to influence speciation in activated sludge plants [11].

Much of the organic matter in wastewater effluents is biologically produced in the treatment process. Understanding of factors controlling production of the material is needed.

It remains to be rigorously demonstrated that introduction of selected organisms or genetically engineered organisms offers advantages in treatment of conventional wastewaters that are not realized by natural selection and evolutionary phenomena. Potential advantages include improved removal rates, reduced synthesis, enhanced nitrification, and improved bioflocculation. Research on the subject must be conducted with greater care than has been typical of previous work. This subject is considered further in the discussion of microcontaminants.

2.1.3 Destruction of pathogenic organisms

Given the increasing incidence of waterborne disease caused by microorganisms and viruses (such as the Norwalk agent) different from those traditionally associated with water, and the high prevalence of water-associated disease in which no causative organism or virus can be identified, it follows that more must be learned about agents of infectious disease in wastewater, and their control. Better indicators of the potential presence of various organisms and viruses are needed. Disinfection practices without environmental effects are needed.

2.1.4 Nutrient removal

Fundamental factors influencing the removal of phosphorus in biological treatment processes remain to be fully elucidated. In view of the attractiveness of avoiding chemical usage producing large sludge volumes, improvement in understanding of removal mechanisms in biological processes warrants attention. Similarly, improvement in economics and control of nitrogen removal is needed.

2.2 Microcontaminants

While the foundation of current wastewater treatment practices rests on methods for removing conventional contaminants considered in the preceding sections, many current water pollution control concerns relate to inorganic and organic constituents in typical wastewaters having potential chronic or acute effects on aquatic, terrestrial, and human populations. These nontraditional wastewater ingredients are typically of concern at concentrations far lower than the concentrations of substances such as suspended solids, biochemical oxygen demand and ammonia nitrogen, and they are referred to here as microcontaminants.

Inorganic microcontaminants of greatest concern are the heavy metals. Organic microcontaminants of greatest concern are anthropogenic. In the United States, the

Environmental Protection Agency's listing of 'priority pollutants' [12] typifies microcontaminants.

In general, research needs related to microcontaminants concern the mechanisms by which they are removed or reasons they are not removed by conventional wastewater treatment processes, necessary modification of conventional practices to achieve removal, fate of microcontaminants in sludge management processes, and changes in those processes needed to minimize the environmental impact of microcontaminants removed by wastewater treatment.

Although this review of research needs does not extend to source control and pretreatment of contributions to municipal sewerage systems, it should be noted that the presence of microcontaminants in municipal wastewaters is an inevitable consequence of modern 'civilization'. Many heavy metals and synthetic organic compounds have become ubiquitous, and even in the absence of industrial sources heavy metals and recalcitrant organic compounds occur in municipal wastewaters from commercial products and activities. Hence, the need for research on microcontaminants in wastewaters cannot be circumvented by regulatory actions.

Given the ubiquity of man-made organic compounds, their presence in municipal wastewaters must be assumed. Because of the ways they are produced, the existence of metabolic pathways for breakdown of the substances in biological wastewater treatment plants cannot be assumed.

Even though anthropogenic organic compounds are fabricated by non-biological means, there is potential for biological removal from wastewaters. The typically low concentration of organic microcontaminants creates problems because the amount of energy available for their biodegradation may not sustain microbial populations and because their low concentrations may not induce the enzymes required for degradation [13]. Schemes for manipulating oligotrophic organisms so as to achieve high degrees of removal of organic microcontaminants are needed. The role of cometabolism in biodegradation of refractory organic compounds must be evaluated and means for exploiting the phenomenon must be developed. Better understanding is needed of the biodegradation of specific trace constituents in complex mixtures of substrates.

Opportunities for selecting microorganisms, for facilitating natural exchange of microbial genetic material, and for human manipulation of microbial genes so as to obtain organisms with unique capabilities for degrading organic microcontaminants must be pursued. Such experiments must be skillfully conducted. As expressed by [14], 'the historic lack of attention paid to the design of such experiments has created a sense of scepticism that should be either substantiated or dispelled: there is no excuse for the accumulation of more meaningless data'.

Adsorption of microcontaminants onto wastewater solids, biological solids, and added adsorbents can be an important means of removal. With slowly degradable organic compounds adsorption offers increased residence time. Improved quantification and control of the mode and role of adsorption is needed. Similarly, the extent of partitioning of some organic microcontaminants into other organic compounds requires exploration. There is a lack of fundamental understanding of sorption reactions in complex solutions containing dissolved organic macromolecules (that may bind both metals and organic pollutants) and a mixture of surfaces including cells, and organic and inorganic particles or coatings. Understanding of the phenomena are important for both fixed-film and suspended growth reactors.

The role of gas transfer in removal of microcontaminants from wastewater needs to be better understood. Means for maximizing or minimizing, as appropriate, the stripping of microcontaminants should be available.

Research on microcontaminants in wastewaters should be conducted in a manner that avoids the historical problems of incorporation of contaminants into sludge without regard to their fate and effects there.

3. SLUDGE TREATMENT, UTILIZATION AND DISPOSAL

The need for sludge management research has become more apparent in recent years as increasing sludge quantities, rising concern for environmental quality control, and escalating costs have made sludge a major problem in water pollution control. The current need for sludge research is intensified by neglect of researchers in the past. Fundamental bases for rational design and operation of the facilities for sludge management are less well developed than is fundamental understanding of the treatment processes that generate the sludge. Sludge management research needs have been analysed recently by a committee of the Water Pollution Control Federation [10] and by Gossett et al. [15].

3.1 Fundamental sludge properties

As pointed out by Gossett et al. [15], if every conceivable fundamental property of sludge were measured, it still would be impossible to predict the performance of conditioning, dewatering and many other sludge management processes. It is necessary to acquire far more extensive knowledge of the basic biological, chemical and physical properties of sludges and to use that information to control sludge properties and to guide the planning, design, and operation of sludge management facilities.

More information is needed on the development of flocculent particles and the factors that influence their density and strength. Although the size of particles in sludges and the distribution of those sizes are known to be important factors controlling sludge behaviour, reliable means for measuring the size distribution of particles as they exist in sludges await development.

Because much of the expense of sludge management is attributable to the water it contains, more must be learned about the manner in which water is associated with sludge solids. In this way, wastewater processes might be operated to minimize the water content of sludges or treatment techniques might be devised to effectively remove moisture from sludges.

The existence of two phases in sludges leads to a need for understanding flow and deformation properties. Improved understanding of sludge rheology is important not only in understanding sludge pumping and piping but also for analysing the physical behaviour of sludges in processes such as thickening and dewatering.

Mechanisms by which potentially toxic constituents such as heavy metals and synthetic organic compounds are incorporated in wastewater sludges must be elucidated. In this way, their inclusion could be minimized as desired and means for extracting the constituents from sludges could be considered.

In summary, detailed understanding of basic physical, chemical and biological properties of sludges must be obtained to serve as a foundation for better understanding, control and economy of sludge management. With such understanding, it should be possible to produce sludges with controlled properties and to rationally develop treatment, utilization and disposal systems.

3.2 Removal of water from sludges

Much of the cost of sludge management is associated with removal of water from sludges or transport and accommodation of water contained in sludges. Moisture is removed by gravity thickening, flotation thickening, dewatering and drying, and conditioning is used to alter the physical properties of sludges to enhance the release of moisture.

Better understanding of the influences of depth, shear, and time on thickener performance is desirable, and this can be achieved by improved basic physical understanding of sludge properties as discussed in a previous section. Improved understanding of flotation thickening requires better understanding of factors controlling association of the solids with gas bubbles.

It is desirable to examine basic physical properties of sludges and then to be able to select and design appropriate dewatering processes based on that information. Much remains to be learned about the dynamics of compressible cake filtration before this will be possible.

Alteration of basic physical properties of sludges by conditioning techniques is necessary before any mechanical means of dewatering currently available are feasible. Ideally, it might be desirable to control wastewater treatment processes so as to generate dewaterable sludges. Realistically, however, reliance must be placed on improved conditioning practices. At present, sludge conditioning is an art form. Research is needed to enable rational selection of efficient conditioning practices based on fundamental understanding of the properties of a particular sludge. This requires development of an understanding of the basic effects of organic and inorganic conditioning agents on particle properties such as hydrophobicity and charge. The effects of various conditioning processes on microcontaminants need exploration.

Currently, there is interest in removing water from sludges by solvent extraction and by multiple effect distillation using an oil carriage system. Conventional heat drying, however, probably does not warrant additional research.

3.3 Stabilization

Many stabilization processes for minimizing odour and other nuisance problems associated with organic sludges are well established. Still, opportunities for improving performance through better understanding of stabilization mechanisms exist.

Better understanding of the effects of aerobic and anaerobic digestion on the physical properties of sludges and their behaviour in subsequent dewatering processes is necessary. More information is needed on the effect of toxic wastewater constituents on performance of biological stabilization processes and on the effect of biological stabilization processes on organic toxic materials.

In recent years, exciting developments in understanding the biochemistry of methane fermentation have occurred, and efficient new configurations of anaerobic biological reactors have been used. Developments based on these advances promise improvement in anaerobic digestion of sludges. Widespread use of aerobic digestion, apart from small wastewater treatment plants, requires economical means of supplying oxygen.

There is evidence of breakdown of some organic microcontaminants under anaerobic conditions while the same compounds resist aerobic biodegradation. Means of exploiting the opportunity require study. Similarly, the fate of microcontaminants in other stabilization processes must be explored.

Sludge composting is an old process that has become popular in recent years. Basic understanding of the process requires research to elaborate mass, energy, moisture and volume balances based on the physics, thermodynamics and biological kinetics of compost systems.

Chemical stabilization of sludges ordinarily has been achieved by using lime to create a high pH to inhibit biological activity. The physical, chemical and biological effects of chemical stabilization and changes in those effects with storage time need to be elaborated.

3.4 Inactivation of organisms and viruses

In recent years, attention has been given to inactivation of organisms and viruses in the United States by such means as pasteurization, high energy electron irradiation, and gamma radiation, and greater interest has been shown in Europe. Perhaps the greatest need for research concerning inactivation of organisms and viruses is to establish the justification for the expenditure. Means for achieving the desired degree of inactivation by various processes must then be explored.

3.5 Thermal sludge treatment processes

Although sludge combustion is an old technology, comparatively recent changes in energy costs have nullified the empirical combustion guidelines that developed during earlier eras. Improved understanding of basic physical properties of sludges and associated improvements in ability to economically condition and dewater sludges will do much to improve opportunities for combustion. Additionally, means must be explored for improving the cost effectiveness of conventional sludge combustion facilities and for evaluating alternatives such as starved air combustion and wet air oxidation.

Efforts to improve understanding of thermal processing of sludges must include research on loss of volatile forms of metals during combustion, destruction of toxic organic compounds, formation of toxic organic compounds during combustion, and control of emissions.

3.6 Reclamation of sludges

Means for recycling sludge constituents into productive uses deserve high priority amongst sludge research needs. It would seem that, ultimately, reclamation strategies must supersede the current focus on ultimate disposal of sludge constituents. Disposal of suitable sludges on to agricultural land currently is the only means by which significant amounts of the constituents of typical municipal sludges are reclaimed. Because of their heterogeneous nature, it is difficult, at present, to conceive of better means for using sludge constituents. Ultimately, reclamation of metals, protein, vitamins, and other trace constituents may render municipal sludges valuable commodities.

Many industrial wastewater treatment sludges are less heterogeneous, and opportunities for reclamation could, with research, be exercised far more commonly than at present.

3.7 Ultimate disposal

If sludge solids can be reclaimed, then there are only three places on earth to which they may be disposed: air, land, and water. While some disposal to air occurs as a result of biological and thermal oxidation processes, disposal opportunities are effectively limited to land and water.

3.7.1 Land application and disposal

Discharge of sludges to land occurs in either a beneficial way (in which sludge constituents are used in agriculture, or silviculture, or for reclamation of low-quality soils), or land is used as a receptacle for sludges.

One of the major potential environmental effects of agricultural use of sludges is the transport of nitrate ions (produced by bacterial oxidation of reduced forms of nitrogen in soils) to groundwaters. Much of the information on nitrification rates has been derived from laboratory studies and more information is needed on the fate of nitrogen in soils modified by the application of sludge.

A major factor controlling the allowable rate of municipal sludges on agricultural land is their content of heavy metals. While many studies have been conducted on the fate of sludge-borne heavy metals in crop plots, the amount of long-term information is small. Only in recent years have plots with an appreciable history of monitored sludge application become available, and they should be monitored closely in the future.

Means of minimizing human food-chain accumulation of toxic constituents of sludges must be explored through consideration of non-food-chain crops and cultivars of food chain crops that will exclude toxic constituents found in sludges.

The fate of refractory organic compounds in soils receiving sewage sludges must be further explored. The accumulation in crops and transport to animals grazing on pasture lands receiving applications of sludges over a long period of time must be investigated.

Municipal sludges offer benefit for reclamation of surface mined lands and other soil of low quality that has not been fully realized. Means for using sludges to convert low quality lands to agriculturally productive lands while reducing problems of erosion and ground and surface water contamination must be further explored.

Landfilling of sludges with adverse properties, while not a satisfying solution to sludge management needs, will continue in the foreseeable future. Chemical, physical and biological transformations that occur under landfilling conditions need to be further explored, as well as interactions of landfill leachates with soils and landfill liners.

3.7.2 Ocean disposal

The evaluation of ocean discharge of sludges has previously been carried out primarily in the public forum without benefit of understanding of the fate of sludge constituents in the ocean environment. The sedimentation, resuspension, solubilization and biological uptake of sludge constituents discharged at sea must be explored. Toxicity of ingredients of sludges to marine systems must be understood. There are signs that mechanisms of pollutant assimilation in the deep ocean are atypical and these must be explored prior to disposal of sludges in these environments.

4. DESIGN AND OPERATION OF WASTEWATER TREATMENT AND SLUDGE MANAGEMENT SYSTEMS

There has been a tendency to optimize the design of individual components of wastewater and sludge management systems rather than to optimize the overall design of integrated systems of municipal processes for wastewater treatment and sludge management. Optimal integration of treatment processes requires a thorough understanding of the effects of design and operational changes on performance. As more is learned about performance of individual processes for wastewater treatment and sludge management, it is to be anticipated that benefits will be derived from improved ability to optimally integrate processes into economical and efficient systems for waste management.

Microelectronic advances have created new opportunities for monitoring and control of wastewater treatment and sludge management facilities. Research must be undertaken to develop a thorough understanding of the behaviour of wastewater treatment and sludge management processes under transient loading conditions. This will permit development of algorithms to allow effective operation under non-steady state conditions. Ultimately, it is to be expected that optimal operational changes will be effected automatically based on rational models of process performance.

5. SUMMARY

The unifying theme of all research suggested in this chapter is that fundamental understanding of basic factors influencing performance of wastewater treatment and sludge processes and of the fate of wastewater constituents in those processes must be achieved.

REFERENCES

[1] Switzenbaum, M. S. (ed.) *Fundamental Research Needs for Water and Wastewater Systems*, National Science Foundation/Association of Environmental Engineering Professors, (1984), 141 pp.

[2] Saunders, F. M. (Chairman), Blaylock, B. G., Boyle, W. C., Braids, O., Campbell, H. J., Jr, Chian, E. S. K., Chu, T-Y., Clesceri, L. S., Clesceri, N. L., Convery, J., Cox, G. V., Gerwert, P. E., Gould, J. P., Grady, C. P. L., Jr, Haasn, C. N., Herricks, E., Hinesly, T. D., Keffer, B., Klemetson, S. L., Meng, H., Miller, D. W., Nelson, R. F., Pohland, F., Porcella, D., Schmidt, J. W., Schroy, J. M., Stover, E. L., Siegrist, T., Throop, W. M., Whipple, G. *Research Needs Associated with Toxic Substances in Wastewater Treatment Systems*, Water Pollution Control Federation, Washington, DC (1982), 299 pp.

[3] Eckert, C. A., Schaeiwitz, J. A., Thomas, E. R. *Research Planning Group Study — Separation Technology. Final Report*, Advanced Environmental Control Technology Research Center, University of Illinois at Urbana-Champaign, Urbana, Illinois, AECTRC Publ. No. 82–3 (1982) pp. 70 (1982).
[4] Johnston, J. B. Robinson, S. G. *Research Planning Task Group Study — Genetic Engineering and the Development of New Pollution Control Technologies*, Advanced Environmental Control Technology Research Center, University of Illinois at Urbana-Champaign, Urbana, Illinois ACETRC Publ. (1983) 83–2 (1983).
[5] Nieman, T. A. *Research Planning Task Group Study — Chemical Detoxification. Final Report*, Advanced Environmental Control Technology Research Center, University of Illinois at Urbana-Champaign, Urbana, Illinois, AECTRC Publ, No. 82–9 50 pp. (1982).
[6] Rittmann, B. E. Kobayashi, H. *Research Planning Task Group Study — Biological Separation. Final Report*, Advanced Environmental Control Technology Research Center, University of Illinois at Urbana-Champaign, Urbana, Illinois, Internal Report, (1982).
[7] Freeman, M. P., Fitzpatrick, *Theory, Practice and Process Principles for Physical Separations*, Engineering Foundation, New York, NY, 750 pp (1081).
[8] Dick, R. I., Sedimentation since Camp, Thomas R. Camp Lecture, *Journal of the Boston Society of Civil Engineers*, **68**, 199–235 (1982).
[9] Dick, R. I. (Chairman), Bova, D., Campbell, H. W., Constable, T. W., Farrell, J. B., Fisher, C.P., Gossett, J. M., Han, G., Haug, R. T., Hinesly, T. D., Jutro, P. R., Novak, J. T., Simmons, D. L., Vesilind, P. A., Wittmer, S. C. and Zenz, D. R. *An analysis of Research Needs Concerning the Treatment, Utilization, and Disposal of Wastewater Treatment Plant Sludges*, Water Pollution Control Federation, Washington, D.C., 111 pp. (1982).
[10] Camp, T. R. A study of the rational design of settling tanks, *Sewage Works Journal*, **8**, 742–758 (1936).
[11] Lee, S. E., Koopman, B. L., Jenkins, D. and Lewis, R. F. The effect of aeration basin configuration on activated sludge bulking at low organic loading, *Water Science and Technology*, **14**, 407–427 (1983)
[12] United States Environmental Protection Agency, Hazardous Waste and Consolidated Permit Regulations. *Federal Register*, 45, 98 (1980), 33066–33588.
[13] Kobayashi, H. and Rittmann, B. E. Microbial removal of hazardous organic compounds, *Environmental Science and Technology*, **16**, 170A–183A (1982).
[14] Johnston, J. B. and Robinson, S. G. *Development of New Pollution Control Technologies: Opportunities and Problems*, Paper presented at 57th Annual Conference of the Water Pollution Control Federation, New Orleans, Loisiana (1984).
[15] Gossett, J. M., Dick, R. I. and Hinesley, T. D. Fundamental research needs for sludge treatment, utilization and disposal. *Proceedings Conference on Fundamental Research Needs for Water and Wastewater Systems* (M. S. Switzenbaum, ed.), National Science Foundation/Association of Environmental Engineering Professors, 104–110 (1984).

4.2

Wastewater treatment with coagulating chemicals

H. H. Hahn
University of Fridericiana, 7500 Karlsruhe, Federal Republic of Germany

1. INTRODUCTION

The present status of wastewater treatment, particularly the area of domestic wastewaters, is characterized by a preponderance of the mechanical-biological treatment concept. Reasons for this are the relative economic efficiency of this combination of unit processes and also the non-specificity, or general applicability, of these processes. However, there is an ever-growing body of literature on the possibilities of extending the limits of efficiency of this process combination to attain the higher degrees of purification demanded by pollution control. These studies indicate indirectly that the boundaries of applicability of this concept are reached in particular in the following areas:

— dissolved inorganic materials are not removed readily or to the degree necessary;
— solids retention has become the needle's eye in the mechanical-biological treatment process in terms of unsatisfactory performance of the secondary clarifier;
— the mechanical-biological treatment plant in its present concept is not suitable for fluctuating or shock loads. (This includes also all load variations resulting from newer concepts of urban drainage treatment.)

Chemical treatment as it is to be discussed here implies the addition of (mostly inorganic) chemicals to the wastewater stream in its course through the existing mechanical-biological plant. These chemicals cause precipitation and coagulation processes, transforming dissolved (precipitating) substances into removable non-dissolved material or improving the liquid–solid separation by increasing the size of the non-dissolved material. In addition there is a considerable amount of adsorption onto the newly formed solid surface. These reactions, however, are envisioned to take place in existing reactors, i.e., in the traditional treatment plant. Thus, the chemical treatment can be accomplished with relatively little investment effort; it causes on the other hand significant operational costs. In many instances the resulting benefits far outweigh the total investment and operational costs.

The following discussion is divided into three parts. In the first, main part the present situation of chemical treatment in terms of technology, efficiency and cost (as well as distribution of this process on a statistical basis) is described. A second part discusses emerging applications for new requirements. The final section contains a summary of unanswered questions that describe future areas of application.

2. PRESENT STATUS OF CHEMICAL TREATMENT IN TERMS OF EFFICIENCY AND COST

The process of aggregation, i.e., formation of larger more easily separable solids from small suspended solids, has traditionally been designed and operated to ensure a maximum efficiency in terms of aggregate growth. Aggregate growth means increase of average particle diameter and therefore better removal in all processes of liquid–solid separation.

In many instances of practical application this goal has been accomplished, but large amounts of solids that cause great difficulties in separation and dewatering have also been obtained. Thus, today the aggregation process will have to be designed and operated such that both 'products' the liquid phase (i.e., the original focus in design and operation) and the solid phase (i.e. the sludge to be expected) will meet certain standards.

There is a great deal of experience and experimental evidence available on the optimization of the aggregation process to obtain a very good clear water quality. This will be presented and interpreted in the following section. The second objective, i.e. to produce manageable amounts of separable and treatable sludge, is presently the aim of several laboratory and technical investigations. The little evidence available for the optimal setting of design and operational parameters to attain this goal is also presented.

2.1 The averge quality of the liquid phase

Aggregation accomplishes the formation of larger non-dissolved solids from small non-dissolved material. It has been pointed out before that in wastewater systems aggregation frequently is initiated by the addition of metal ions (metal salts). Thus, depending upon the composition of the dissolved phase, there will be more or less pronounced precipitation of substances that form insoluble complexes with the added metal ion (for instance Me-phosphate).

When discussing the efficiency of the process in terms of clear water quality, one has to bear in mind that only aggregating and precipitating substances will be affected. It has also been indicated above that there is adsorption on to the freshly formed solid surface when aggregation and precipitation processes occur. Thus the concentration of adsorbing substances will also be affected by this process. At the same time it must be emphasized that none of the constituents of the dissolved phase of a wastewater system will be removed or altered if those substances are not amenable to precipitation and/or adsorption. Nitrogen species are a case in point. They will not be removed in any instance of chemical dosing that has been described here.

It is important to accept that any quantitative information on removal efficiencies or on effluent quality is, and must be, problem specific. There are so many interfering reactions for an unknown or un-investigated wastewater system that it is difficult to predict the type and extent of all processes that might occur. A general listing of process efficiency data that might be desirable for the design engineer is not possible. However, one can inspect efficiency data reported in the literature, evaluate and discuss them within the context of the specific situation, and derive from this expected orders of magnitude of the process efficiency.

The process is monitored in agreement with the above-described principles in terms of turbidity reduction and/or reduction in filter residue; decrease in the concentration of specific ions (e.g. phosphate, heavy metals), and decrease in biochemical/chemical oxygen demand (for a filtered or unfiltered sample). These are certainly not all the parameters that one would select on the basis of known process efficiencies. However, routine analysis during (treatment) plant operation usually does not allow more specific investigations, such as change in particle size distribution (very specific for the description of this process), or reduction in specific organic substances.

Furthermore from a practical point of view such parameters should be listed in more than one 'dimension', for instance as: (a) first statistical moment (mean, etc.) of distribution of effluent concentrations, and/or (b) second statistical moment (standard deviation etc.) of distribution of effluent concentration, and/or (c) similar measures for (relative) concentration reduction in the effluent. Each measure will describe a different aspect of process efficiency and operation. And each might be of particular importance under different conditions.

2.2 Actual plant efficiency data as reported from process operation

As indicated above, the use of precipitating or coagulating chemicals is always recommended when the following wastewater constituents are to be controlled: (a) non-dissolved suspended solids which are registered in the parameter turbidity or filter residue; (b) non-dissolved or dissolved adsorbing organic substances which are registered in the parameter BOD resp. COD; (c) dissolved inorganic substances which form precipitates with the metal ion.

This practical limitation of the process to the control of only some wastewater constituents is shown in Fig. 1 which also indicates that there are always several processes occurring at the same time when metal ions (or coagulants) are added to a wastewater system. It depends upon the relative preponderance of the specific wastewater constituents which specific reaction pathway is favoured.

When discussing process efficiency in real systems one must realize that coagulation/flocculation and the competing precipitation process depend for their success upon the subsequent step of liquid–solid separation. Thus, in all discussions it is assumed that there is an optimal process of solids separation available.

Figure 2 [1] shows a collection of data on the effectiveness of the dosing technical iron or aluminium salts into the mechanical or biological stage of traditional treatment plants (i.e. preprecipitation or simultaneous precipitation). Two phenomena should be pointed out specifically. (a) There is a quantitative relationship between chemical dosage and process efficiency, i.e., with increasing dosage the efficiency rises within certain boundaries. (b) The relationship between chemical

Ch. 4.2] Wastewater treatment with coagulating chemicals 147

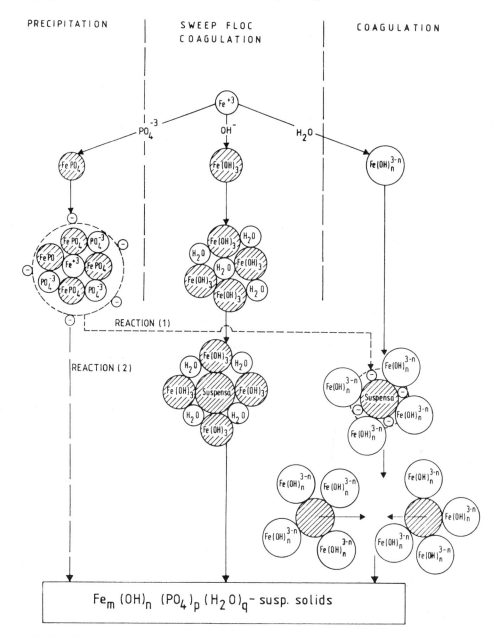

Fig. 1 — Simultaneous processes occurring when metal salts are introduced into a wastewater stream for reasons of coagulation and precipitation.

dosage and process efficiency as defined by practical observations is not a uniquely defined curve but rather a domain.

In order to be able to compare this diagram with other, more fundamental

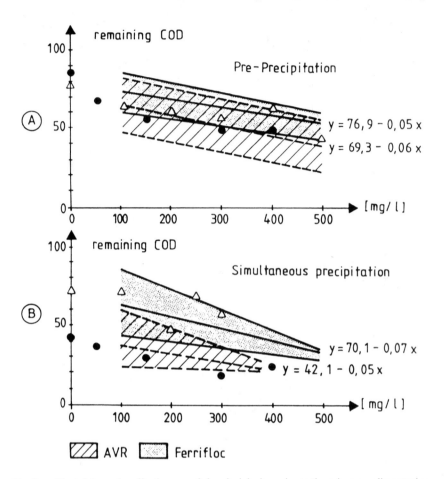

Fig. 2 — Plant data on the effectiveness of chemical dosing prior to the primary sedimentation tank and prior to the secondary sedimentation tank. The effect of increasing chemical dosages is readily seen.

information on the efficiency of chemical dosing one must be aware that technical coagulants used contain only about 10 per cent active material. Other chemicals, such as calcium or polymeric substances require higher (in the case of calcium) or lower (in the case of polymers) dosage.

The comparison of plant efficiency of a mechanical-biological plant built in the standard of 'generally available technology' with efficiency data for such a mechanical-biological plant that are supported by chemical dosing is given in Fig. 3 [2], [3], [4], [5], [6]. It is very clear that the addition of chemicals leads to a significant increase in the quality of the performance.

In addition to the absolute increase in the removal or reduction rate of undesirable wastewater constituents there is also an increase in the stability of the performance, i.e., a reduction in the scattering of the efficiency data when chemicals are used. Cumulative frequency distributions obtained for plants with chemical

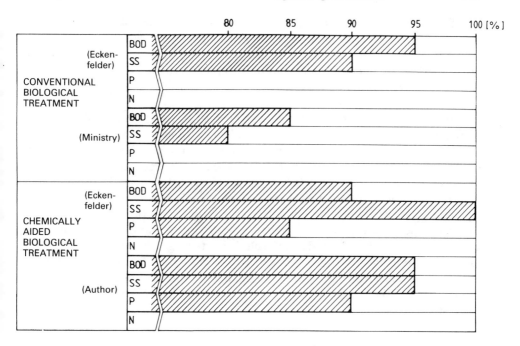

Fig. 3 — Comparison of various treatment concepts in terms of their efficiency in removing wastewater constituents. The figure shows the relative increase in removal efficiency upon the dosing of chemicals.

dosing show a much steeper line, i.e., a lower standard deviation than those lines obtained or observed with mechanical biological treatment alone.

The observed improvement of plant performance, i.e., the reduction in fluctuations of the effluent concentrations results from two phenomena: (a) inflow fluctuations are dampened by the frequently flow-proportional operated chemical dosing; (b) the process of precipitation/coagulation allows a rather rapid response to known or anticipated load fluctuations such as from stormwater runoff etc.

The full potential of stabilizing or equalizing the plant effluent has not yet been realized. The automation of this particular process is, at best, at its beginning. Contrary to traditional biological processes this chemical process is described and controlled by analyses that are readily and rapidly feasible. There is already some experience in such plant control. Finally, all other reactions in the course of the wastewater treatment process will profit from the addition of a treatment step that leads to load reduction and to load fluctuations. Thus, the overall plant performance of such plants where chemicals are used for precipitation/coagulation is more than linerally improved.

3. THE COST OF CHEMICAL WASTEWATER TREATMENT

The decision for or against the use of a (technical) process is always based on two aspects (a) the efficiency of the process or its potential, and (b) the cost of the installation and operation of the process.

The efficiency of the process has been discussed in terms of reaction rate and reaction end-point in the preceding section. Many of the deciding factors in the practical potential of this process, such as physical parameters (energy input, retention time, geometric proportions of the reaction chamber, etc.) and chemical parameters (type and amount of chemical, etc.) can be identified and described quantitatively.

Cost figures for wastewater treatment processes are difficult to obtain. Thus, there is reason to hesitate in applying cost estimates from one project to another; (a) while the process might be the same the surrounding conditions are frequently significantly different; (b) cost figures depend to a large degree upon the loading of the respective plant, i.e., the specific wastewater characteristics — and those are, in almost all instances, different from application to application; (c) the relative size of a plant has a marked influence upon the specific cost figures (for instance expressed as unit cost per unit water treated), i.e. the so-called 'economy of scale' effect, which again makes a transfer of cost data difficult. Nevertheless, it is necessary to have cost estimates as reliable as possible for a given wastewater treatment process before the decision for its inclusion into the overall treatment process is made. It is within this framework that the following remarks on the cost of 'chemical treatment' are to be understood.

3.1 Cost of chemicals, chemical storage and dosing

In estimating costs for a process it is important to identify where and in what form (investment cost, operating costs, etc.) costs might arise. If one analyses a project then one finds that investment costs occur to a significant degree only when 'post-precipitation' or 'post'-coagulation is applied (most frequently in the context of phosphorus removal in domestic treatment plants). In this instance the coagulation reactor and the additional plant for liquid–solid separation, i.e., frequently a flotation unit, are responsible for these investment costs. In the case of 'simultaneous' dosing of chemicals, installations are only needed for the storage of the chemicals, their dosing and the mixing. In the first approximation one can neglect these smaller investment costs and treat 'pre'- and 'simultaneous' coagulation as operating-cost intensive unit processes. This means that costs occur only when the process is in operation. (This fact, along with the observation that the process can be started very quickly, makes it very attractive for the reduction of load fluctuations.)

If one compares the individual cost factors one finds that, aside from the expenses for sludge handling and sludge disposal (discussed in Section 3.3) the process costs are determined by the chemical costs. In direct relation to this the management of chemical storage will affect the overall costs. Chemical costs in themselves are proportional to the amount of chemicals used. If one multiplies, for instance, the abscissa of Fig. 2, i.e., the necessary doses of precipitating or coagulating chemicals, by the unit price of those chemicals then one obtains the total cost as a function of the efficiency of the process. The result of such a cost calculation is shown in Fig. 4 [5] for some commercially available chemicals (in the Federal Republic of Germany). It is interesting to note that there is very little economy of scale displayed by the cost function. This results from the dominance of the water throughput-dependent chemical costs.

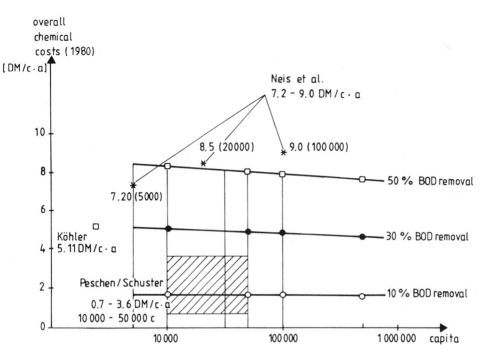

Fig. 4 — Cost estimate for chemical treatment (primarily based on the cost of purchase and storage of chemicals).

Other cost factors that might have to be considered, particularly if the process is included at the end of a mechanical–biological treatment plant, are costs for energy input and also costs for the units process of liquid–solid separation.

3.2 Cost of liquid–solid separation after addition of chemical i.e. flotation

If, in the case of 'post' treatment, an additional liquid–solid separation step is needed, then flotation will be advisable for reasons of (1) better clear water quality, (2) better response to voluminous flocs resulting from Me-salt addition, and (3) higher solids content in the sludge. In this instance, additional installations are needed. Therefore there will be investment costs. In a similar way the operating, maintenance and repair costs must be estimated.

For a comparison of different procesess, in particular an operating-cost effective one and one that is significantly determined by investment costs one conventionally uses total costs, i.e., the sum of investment and OMR. They can be found by either adding up all OMR costs over the total lifetime of the process or by taking the annual amortization and interest of the investment costs and adding them to the OMR costs.

In Fig. 5 the total costs are given as specific or unit costs, relating them to the amount of wastewater treated (and taking into account either the conversion of investment costs into annual costs or the necessary summation of the OMR cost over the whole lifetime of the unit process). The specific cost figures show a noticeable 'economy of scale'. This results from the necessary installations.

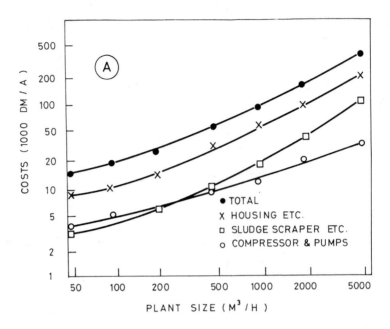
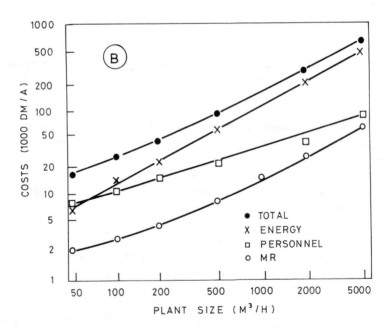

Fig. 5 — Cost of flotation as one of the effective liquid–solid separation processes used in combination with the addition of chemicals within the context of mechanical-biological treatment.

3.3 Cost of handling and depositing of (extra) sludge resulting from the use of chemicals in wastewater treatment

If the cost of sludge handling and deposit is to be estimated then the sequence of treatment steps needed must be known. For the purpose of this cost estimate two alternative ways of sludge treatment were envisioned [7]: (a) joint treatment with (predominantly organic) sludge from standard domestic treatment plants, and (b) separate treatment in specifically designed sludge handling plants that take into account the possibility that this sludge has a higher content of inorganic material, i.e., must not necessarily be stabilized. For these types of sludge handling plants total annual costs had to be determined. The necessary input data required for the design of such plants with respect to solids production, solids content etc. have been taken from the literature.

The type of treatment recommended in this investigation for combined domestic mechanical-biological sludge and chemical sludge is indicated as: pre-thickening, anaerobic stabilization, post-thickening, chemical conditioning, filter press dewatering, transport, deposit in controlled landfills. This appears to be the most economical type of treatment for larger installations while in the case of smaller plants agricultural use might be possible. Then no dewatering, or a less costly dewatering process, might be sufficient, generating lower costs.

A second alternative for the treatment of sludge from chemical wastewater treatment assumes that the organic content of the sludge is smaller. Therefore no stabilization is required. Furthermore, it is assumed that this sludge will not, or cannot, be deposited onto agricultural areas (for instance, due to a high content of heavy metals not necessarily connected with the use of chemicals). Again, as in the case of the situation described above, the least cost solution is developed through an optimization routine. And again there will be other routes of treatment and disposal for smaller installations possibly leading to somewhat lower costs. The line of treatment consists of: (1) prethickening, (2) chemical conditioning, (3) filter press dewatering, (4) transport and (5) landfill. The results of these cost calculations are shown in the following section.

3.4 Comparison of cost of chemical treatment with the cost of other treatment methods

As has been pointed out several times, it is not intended that the cost data given in the preceding paragraphs are used for actual design purposes; they should rather serve as an illustration. They should also show in what direction costs will develop if certain process specifications are changed. One can, however, exploit the meaning of these cost data to a larger extent if one compares them with the cost of other unit processes used in advanced wastewater treatment or even with the cost of the well known mechanical-biological wastewater treatment. In Fig. 6 [8] the total (unit) costs for various treatment concepts are given. In order to show the effect of the 'economy of scale', i.e., the size of the installation, the cost figures have been developed for two significantly different plant types (a) type I: 10 000 capita; (b) type II: 100 000 capita. The change in cost as a result of increased plant size is seen clearly.

A comparison of the various cost data given in Fig. 6 shows the following: (1) mechanical-biological treatment (as the basis for today's treatment concept for predominantly domestic sewage) is costly when compared with the other unit

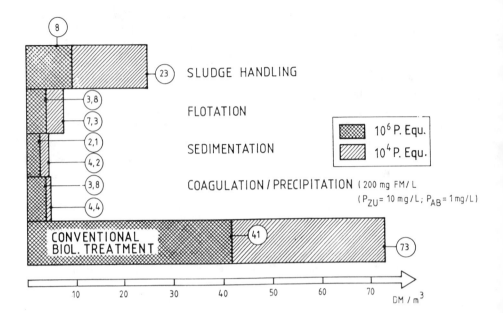

Fig. 6 — Total unit cost for various unit processes for two plant sizes.

processes; (2) chemicals (addition) for precipitation and coagulation is less expensive by nearly an order of magnitude than mechanical-biological treatment; (3) in comparing the effect of the economy of scale one notes a much more pronounced influence upon the process of mechanical-biological treatment; (4) sedimentation, flotation and filtration as processes of liquid—solid separation are all characterized by comparable cost figures (i.e., the cost differences are small when compared to the dimensions of the cost of mechanical-biological treatment; (5) sludge treatment (without the cost of disposal) is more expensive than physico-chemical treatment; (6) in the case of sludge treatment the phenomenon of cost decrease is clearly noticeable; (7) for larger installations the sum total of chemicals addition and liquid–solid separation is comparable in magnitude to the expense of the necessary additional sludge treatment.

In summary one can conclude that 'chemical treatment' is not only a process that is well proven in terms of its efficiency. Its cost can also be estimated and will, under most conditions, be significantly lower than the cost of mechanical-biological treatment. This is not an argument for substitution of those traditional treatment plants by chemical treatment plants; it indicates, however, the meaningful application of this process in reducing load fluctuations.

4. STATISTICAL EVALUATION OF THE DISTRIBUTION OF THE PROCESS

The Federal Republic of Germany may represent average conditions with respect to wastewater composition and efforts to treat this wastewater in the context of industrial country standards. Thus it might be interesting to analyse the present-day distribution of this process and to ask what the reasons for its application (or lack thereof) are.

At present (end of 1983) chemical coagulation/precipitation is applied in some 150 municipal sewage works in the Federal Republic of Germany for intensified or advanced wastewater treatment. Re-calculated this represents some 8% of the total domestic sewage treatment capacity (corresponding to roughly 9 million population equivalents; see ref [9]).

Plant operators' interviews disclosed a number of reasons for using chemicals within the conventional treatment process: increased phosphorus removal from primary and secondary effluents; improved removal of suspended solids and correspondingly better plant performance in terms of BOD and COD removal; control of bulking sludge and improvement of overloaded plants. It is interesting to note that the results of the questioning shown in Fig. 7 identify phosphorus removal as the main reason for the aplplication of chemicals.

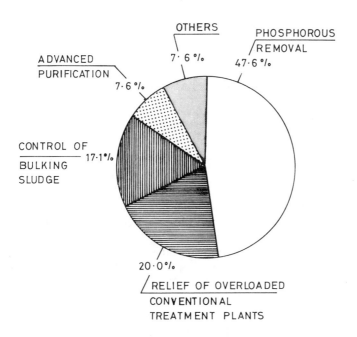

Fig. 7 — Main reasons for chemical coagulation and precipitation within mechanical-biological treatment plants in the Federal Republic of Germany (results of a statistical evaluation).

The dosing of chemicals within the conventional treatment process occurs in many different ways in these plants, depending upon the physical layout of the plant and also on the specific requirements of the receiving water. If one distinguishes roughly 'primary' or pre-coagulation, 'secondary' or simultaneous coagulation and 'tertiary' or post-coagulation, then one finds that close to 50% of all installations use the simultaneous coagulation approach (see Fig. 8). Similarly the inquiry showed that close to one quarter of all plants do not adhere to these standard practices and use 'specific' process variations. The latter include dosing at more than one point, as well as the use of chemicals only (without the supporting biological process).

Finally, the survey disclosed interesting information on the amount of sludge produced by the addition of chemicals in mechanical-biological plants. This question has frequently been overlooked and has led to grave difficulties in those cases where the actual sludge volume increased unexpectedly. Sludge production from wastewater treatment is influenced by a number of variables such as characteristics of the raw sewage, intensity and type of treatment and handling of the sludge. When chemicals are added additional matter is removed from the wastewater stream and therefore the sludge production enhanced. Furthermore, these chemicals frequently show a high affinity to water, thus leading to increased water content of the sludge.

While the literature shows conflicting information on sludge production (up to 450% increase for primary coagulation, 38% less to 160% more for secondary coagulation, and 60 to 100% more for tertiary coagulation), this study identified that on a statistical basis the additional sludge volume produced is within 30% of the total sludge of the conventional plant.

5. AVAILABLE EXPERIENCE IN TERMS OF PLANNING, CONSTRUCTION AND OPERATION OF THIS PROCESS

5.1 Operational aspects: chemicals selection

In many instances the reactors for liquid–solid separation do exist and the overall plant efficiency can be increased by an improvement of the 'separability' of the solids, i.e., inducing aggregation of particles [10]. The most important decisions in such instances are (a) point of chemicals addition (b) type and dosage of chemical. For this reason the selection of chemicals deserves great interest, be it for the optimizaiton of the clear water quality or of the sludge characteristics or for reasons of operational reliability and robustness.

As will become clear from the following, the various chemicals available to the operator of the wastewater treatment plant have significantly different effects and also significantly different consequences for the overall purification process. Therefore it is necessary to describe briefly each of the chemical types used today.

As mentioned before these are basically inorganic (mostly metal salt-type chemicals) and organic (mostly polymeric chemicals used in today's water technology [11]. The differences in reactivity of coagulating chemicals can be described as follows.

Calcium: Effects: coagulation due to counter-ion effect (precipitation of calcium is negligible). Remarks: no problems with over-dosing, high amounts of chemicals needed (Fig. 11), higher pH values used in actual operation, larger amounts of solids (sludge) produced, dewatering of these solids not too problematic.

Fe III/Al: Effects: counter-ion coagulation at lower pH values and surface-charge reduction (coagulation) at higher pH values of about 5–7. At even higher pH values hydroxide precipitation will occur. Remarks: at lower pH values the necessary dosage is relatively low, at intermediate pH values the system is very pH sensitive but

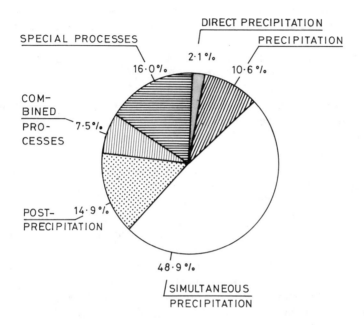

Fig. 8 — Different chemical dosing points or different technologies for chemical treatment in the context of German municipal treatment plants (results of an inquiry).

very effective in terms of necessary dosing, at higher pH values large amounts of chemicals are needed for the precipitate formation. Over-dosing, in the sense of decreasing effectivity, at increased dosage rates, becomes a problem when the system is to be operated at intermediate pH values. The amount of solids forming and the problems in dewatering the resulting 'sludge' are large when the system is operated at the hydroxide formation stage while in all other conditions these aspects are non-problematic. Fe^{2+} is a less efficient coagulant; however, if oxidized or oxidizable it is an economic one.

Polyaluminium: Effects: change of surface charge through adsorption of highly charged low molecular weight hydroxo complexes; Remarks: conditions for application as above for Fe III/Al at intermediate pH values, pH sensitivity less pronounced.

Polymers — cationic lower molecule weight: Effects: modification of surface charge through adsorption of material; Remarks: close similarity to (inorganic) polyaluminium in dosing requirements and effects.

Polymers — cationic, high molecular weight: Effects: bridging through adsorption of long-chain molecules at more than one particle surface; Remarks: very low dosage requirements, the pH regime must be closely controlled; when over-dosing occurs then the (changed) surface charge is reversed and restabilization begins, aggregates formed are voluminous and frequently show unsatisfactory dewatering properties.

Polymers — anionic, high molecular weight: Effects: bridging between particles that frequently carry a charge of the same sign, intermediary reactions with other constituents of the dissolved phase is assumed (for instance with Ca^{2+}); remarks: very low dosage requirements, the pH regime need not be controlled as closely as described above as charge variations do not have such effects, similarly dosing need not be controlled as accurately as above, solids and sludge characteristics are mentioned above.

Polymers — non-ionic, high molecular weight: Effects: bridging between particles (see above); Remarks; pH effects completely disappeared (for practical purposes), all other conditions as above.

The type of chemicals selected will also affect the strength of the floc formed. Furthermore, the phenomena of simultaneous coagulation and precipitation will influence the overall results. Coagulation is the formation of larger aggregates from solid substances, i.e., no change in phase. Precipitation, the formation of solid, non-dissolved species, implies a phase transition. In chemical terms coagulation and precipitation are distinctly different processes. In wastewater systems, however, in particular when metal ions (metal salts as coagulants) are used, both processes might occur simultaneously. To what extent these two different processes occur and with what reaction rate they proceed depends upon the composition of the dissolved phase. This dissolved phase is very complex and changes in nature in wastewater systems. In particular in the presence of phosphate ions, for instance, precipitation will prevail. If, for instance, the hydroxide ions predominate, then metal hydroxides will be primarily formed. Under conditions of intermediate pH values dissolved hydroxo-complexes will be formed leading to coagulation. Chemical addition in the form discussed here leads by definition only to a change of phase or to aggregation/floc formation. It does not allow *per se* the elimination or modification of substances that do not precipitate or coagulate. However, in practical applications it has been observed that constituents of the dissolved phase that do not belong to either of the above-mentioned categories are also reduced in concentration. The most plausible explanation is a sorption process which is favoured by the very large specific surfaces formed for instance in metal hydroxide precipitation [12], [13]. It is reasonable, therefore, to expect such adsorption processes to occur in situations where solid surfaces are formed or reformed and where surface active species are available (as in the case in wastewater). Examples of such elimination by adsorption on to coagulated solids are heavy metals removal or DOC reduction in wastewater systems treated with coagulants. This also has to be borne in mind when chemicals are selected.

5.2 Planning and construction: reactor design

For practical design (and operation) the following facts have to be kept in mind. The retention time of a reactor is affected by: (a) flow rate relative to volume; (b) energy dissipation in the reactor (relative to the energy introduced with the throughput); (c) stirrer type and geometry (relative to the reaction chamber); (d) inflow and outflow configuration (relative to the geometry of the reactor); (e) compartmentalization of the reactor; (f) baffles and other flow directing devices in the reactor (relative to the reactor geometry). In the past these parameters have not been controlled closely or extensively, whether in the phase of reactor design or for the repeated optimization during operation. This is true for both types of application, aggregation reactors and flotation reactors. Frequently, a 'sweeping' effect of chemicals used in the aggregation reaction and the flotation reaction has masked possible problems resulting from non-optimal design (and operation). Yet, presently it appears necessary and also possible to exploit reactor dynamics more extensively in order to save chemicals or to attain higher degrees of efficiency.

If coagulation/flocculation occurs primarily due to velocity gradients, as is the case in most technical reaction chambers, then the effect of non-homogeneous energy dissipation/velocity gradients must be investigated. It has been shown that the reaction progress for a specified reaction time proceeds directly with increased energy input or energy dissipation. A specific energy input into a reactor of given volume might lead to a homogeneous pattern of velocity gradients. Then the above rate law will describe adequately the reaction progress. If, however, the geometry of the reactor system is such that there are zones with very high energy dissipation rates and other zones with rather low dissipation (still leading to the same average velocity gradient as in the first case) then the resultant overall net reaction rate will be lower. Here one could calculate the effect of such heterogeneity by 'decomposing' conceptually the reactor in 'n' sub-reactors with quasi-homogeneous energy dissipation.

Laboratory and plant data show clearly that there are significant differences between different stirrer and reaction chamber configurations. Despite identical calculated detention time and energy dissipation there are noticeable differences between the coagulation rates. In particular (a) different stirrers show different sequences of effectivity in different reactors (b) if stirrer 'a' is better than stirrer 'b' in the jar reactor then this must not be true for the continuous flow reactor (c) if stirrer 'a' is better then stirrer 'b' in the one-compartment reactor than this must not be so for the two-compartment reactor.

From these arguments the following recommendations derive for the stepwise design of coagulation/flocculation reactors: (1) The design objective is, for instance, a reactor that converts a known suspension (i.e., of known average particle diameter) into one with a larger diameter in order to guarantee removal by a liquid-solid separation process as flotation. (2) In the next step experiments of an exploratory nature are to be performed which should show the possibilities of aggregating the suspension (and the necessary type and amount of chemicals). (3) Then from the 'idealized' rate law described an estimate is made for the necessary detention time and the required power input. Both parameters must only be estimated within certain ranges. They are also interdependent: a higher energy input will lead to a lower necessary reaction time and vice versa. Estimated values for these parameters can also be taken from the literature. (4) Now the reactor geometry has to be determined (after the overall size has been set by the determination of the detention or reaction time). This should (and can only) be done by 'scale-model' experiments. By such experiments all effects discussed above will be evaluated *in toto*. Possible systems to be investigated should be. (5) It is important to point out that the scale-up of these models is difficult and critical and that there exist no rules for this. Depending upon the situation one can choose between various dimensionless or characteristic numbers, such as the Froude number or the Reynolds number (for the reactor or the stirrer) or a power input related scale-up number [14]. (6). Finally it must be stated (again) that jar-test-type experiments on the potential for aggregation of the suspension should be used to optimize the operation. Such optimization will be necessary again and again if the characteristics of the influent suspension will change. The analysis can also help to overcome possible shortcomings of the design.

6. CHEMICAL TREATMENT FOR THE CONTROL OF FLOCCULATION LOADS

6.1 The problem

Fluctuating loads lead to noticeable negative effects on the performance of (conventional) mechanical-biological treatment plants. Such load fluctuations may result from the routing of combined sewer discharge through the treatment plant. It may also result from industrial discharges, in particular in the realm of seasonally oriented food industries. Finally the effects of ever-growing tourism are also reflected in intensified time-variations of wastewater production. One possibility for the control of such peak loads is storage in specific reactors which as a rule should be aerated. Such storage is only advisable if the stored waste does not show too rapid decay and if the ratio of peak load to base load is not too large; otherwise the necessary storage volume is unattractive from an economic point of view (see Fig. 9).

Biological treatment processes can follow such fluctuations only within limits. The response time of such controls is effectively limited by the amount and consistency of return sludge that can be re-introduced into the activated sludge tank. The control of highly varying loading peaks by chemical means has been proposed in various instances and also shown feasibility in some particular situations, such as the treatment of vinery sewages etc. [8]. The literature contains convincing arguments that the cost of this additional chemical treatment competes under such circumstances successfully with the cost of conventional mechanical-biological treatment.

6.2 Characteristics of peak loads and response of chemical treatment (including liquid–solid separation)

The amplitude and frequency of load fluctuations depends upon the origin of the waste stream. Typical frequencies of precipitation caused by peak loads are in the order of magnitude of days, with amplitudes (by definition) a factor of two larger than the base load. On the other hand load fluctuations resulting from food industries, such as vine production have peaks of close to ten times the magnitude of the base flow and durations of weeks to months. An analysis of these different phenomena shows also that in particular those peaks that have a high maximum (and

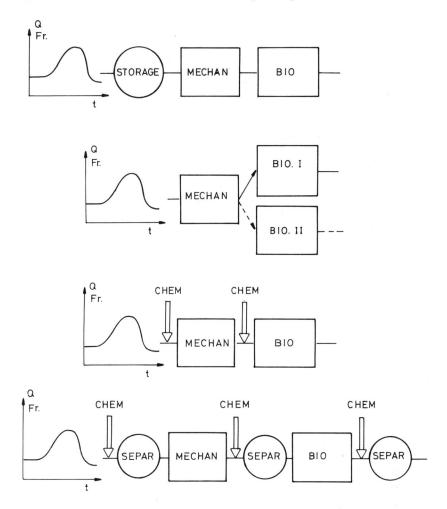

Fig. 9 — Schematic drawing of the different possibilities for the reduction of load fluctuations, indicating also the conceivable points of chemicals dosing.

frequently a lower return interval) are amenable to chemical treatment. It is to be concluded from this that tourism-caused load variations and agro-industry discharges can and should be controlled with chemical addition within the mechanical-biological process.

Chemical reactors are expediently operated as completely mixed reactors. They should be operated in agreement with the principles of such detention time behaviour. The subsequent liquid–solid separation chambers, be it sedimentation or flotation units, are designed to perform as plug flow reactors. From this analysis one can then predict the response characteristics to changes in the reactor performance on changes in the operating conditions, i.e., the response of the effluent concentration on changes in the inflow concentration.

It can be described by analytical means that a plug flow reactor responds faster to changes in inflow concentration variation and that the completely mixed reactor requires more than two to three detention time periods to reflect to a satisfactory degree the new equilibrium. Thus the overall response time of a 10–20 min detention time coagulation reactor and a 0.5–2 h detention time separation reactor is so short that the above described load fluctuations can be controlled.

The addition of chemicals can be done at various points of the (existing) mechanical-biological plant (see Fig. 9). Decisive in the choice of point is the type of chemical used and the access to the plant element. Besides these arguments it has to be decided how the dosing is to be programmed (mechanically according to predetermined schedules, or automatically following identified inflow characteristics). There are several practical possibilities today, such as: (a) constant dosage; (b) predetermined load characteristics as control; (c) discharge proportional; (d) real-load proportional. Without doubt the last mentioned strategy is to be preferred for this particular task. However, the possibilities of measuring specific wastewater constituents in real time in the day-to-day operation of a treatment plant are limited to possible turbidity and phosphate measurements (for the purposes discussed here). If calcium is used for precipitation coagulation then a pH-control may be installed for efficient dosing of the chemical.

In general, however, there are only very few examples of efficient automated dosing of chemicals in wastewater treatment plants. It is certainly one of the areas where further (applied) research is needed if this type of treatment is to be exploited further.

6.3 Examples of successful application of chemical treatment for the control of load fluctuations

It is neither necessary nor possible to present in this context all conceivable examples of successful chemical applications. The selection of the two cases to be presented here is based on the notion of illustrating principles and of presenting convincing material. Furthermore, the case studies should represent instances of 'generally available or acceptable technology', i.e., they should demonstrate a general feasibility.

Example 1 deals with the pretreatment of an (industrial) waste-stream prior to the mechanical-biological plant that is operated by the nearby community. The time-variant load peaks from this discharger cause a frequent break-down of the central treatment plant. The industrial operation can be classified as food industry, producing large amounts of readily biodegradable dissolved and non-dissolved carbonaceous organic material.

The installed pretreatment consists of a dosing station arranged in the 'sewer' i.e., a tubular reactor. Here inorganic coagulant is added (30–60 mg/l of Fe III) supported by a subsequent addition of flocculant aid (cationic high molecular weight material in the order of 1–10 mg/l). The chemical dosing is followed by a high-load flotation unit of the dissolved air type.

Figure 10 shows a summary of data obtained in a large-scale application of this process. The upper part of this figure shows inflow and effluent concentration of total suspended solids, while the lower part of this figure contains the observed effectivity on the (filtered) COD reduction. The inflow data of both wastewater constituents

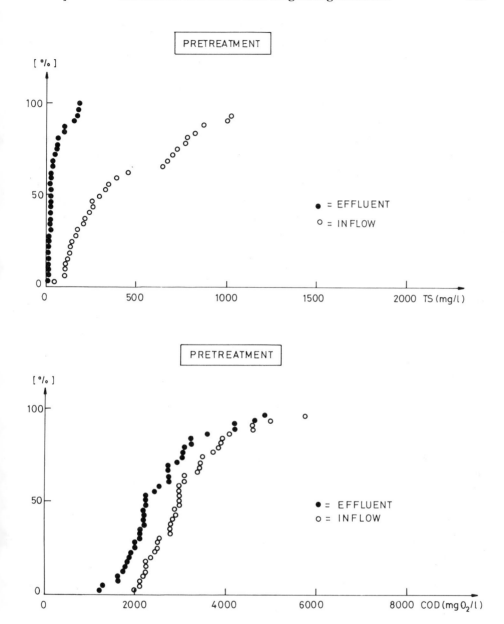

Fig. 10 — Inflow and effluent concentration of total suspended solids and COD for a pretreatment step by means of chemicals addition into an industrial waste stream prior to a municipal treatment plant.

show highly variable characteristics (indicated by the relatively shallow slope of the cumulative frequency distribution). The effluent concentration of the suspended solids fraction is significantly reduced (relative to the inflow) in terms of mean concentration value and, even more significant in this context, in its variability (steep slope of the frequency distribution). Contrary to this the effluent concentration of the (dissolved) organic fraction is not effectively controlled, neither in terms of mean concentration values nor in terms of variability. This, incidentally, illustrates clearly the possibilities and limits of the so-called chemical treatment (see above).

Example 2 also illustrates the problems of industrial discharges into municipal sewers and municipal treatment plants. Here the solution of the problem is conceived in a different way, partly because there exists more than one discharger. The support of the mechanical-biological treatment plant consists of a tertiary stage where chemicals are dosed into the effluent of the conventional secondary sedimentation tank. The liquid–solid separation is accomplished again by a high load flotation unit of the dissolved air type. One might claim that this is the other end of the spectrum of relief or support measures for treatment plants with problems resulting from fluctuating loads.

Figure 11 shows again inflow and effluent data for the two characteristic wastewater constituents, total suspended solids and (total) organic material assessed as COD. From the data plotted as frequency distributions it becomes clear that the variability of the wastewater composition is very high, i.e., load fluctuations are significant. The effluent concentrations of both wastewater constituents are significantly reduced in terms of mean concentration and variability as well. Again the precipitation coagulation process proves effective for the reduction of total suspended solids. In addition the organic load of the effluent stream is significantly reduced since most of it appears to be in the form of (biological) flocs.

In summary then one can conclude that the problem of load fluctuations can be effectively solved by applying precipitating and coagulating chemicals (followed by a liquid–solid separation, frequently installed as additional plant), in particular when the problematic wastewater constituents are in a form where they can be coagulated, or precipitated or adsorbed.

7. FUTURE DEVELOPMENTS NEEDED FOR IMPROVED APPLICABILITY OF CHEMICAL TREATMENT

7.1 Automatic control

From a practical point of view, wastewater treatment must be accomplished in the near future by processes that can be automated or that are already 'self-controlled'. Reasons for this are, on one hand, the high costs of personnel (as well as the difficulties in finding sufficient qualified specialists for the complex tasks at hand), and, on the other hand, higher and higher expectations with respect to the quality of the output of the 'industrial process called wastewater treatment'. Such automation calls for instantaneous assessment of the situation of the process and also for a fast response to the process to any external manipulation.

One of the weak points of the conventional mechanical-biological process is the difficulties in controlling this process effectively; this stems on one hand from the lack of analytical instruments that assess the quality of the incoming (and possibly outgoing) waste stream and, as indicated above, from the slowness of the response of the process. 'Chemical' treatment is characterized by a considerably faster response, as shown before. Furthermore, the crucial parameters for assessment of the process

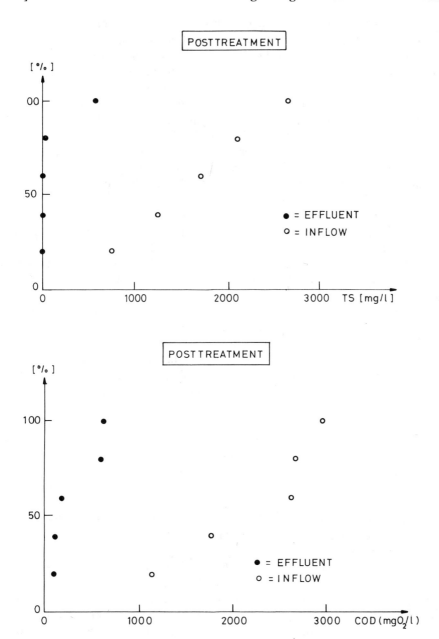

Fig. 11 — Inflow and effluent concentration of total suspended solids and COD of a tertiary treatment step following a conventional treatment plant that is overloaded with peak or fluctuating loads. The treatment consists of chemicals dosing and liquid–solid separation.

efficiency can be determined continuously or semicontinuously under laboratory conditions (such as pH or turbidity, or phosphate).

The statement that all necessary elements of an automatically operated chemical treatment step exist and are proven individually in the laboratory indicates the future developmental needs. First of all the analytical instruments need to be evaluated extensively in the field, i.e. in the day-to-day operation of municipal wastewater treatment plant. 'Drift' problems, etc., should illustrate the kind of questions to be asked and answered. (This, however, is a question that is not central to this particular discipline.) Then the coupling of individually tested elements, for instance their positive or negative feed-back relations, must be explored possibly in pilot plant and large-scale technical experiments. Without doubt one can predict the result of superposition of the different elements on an analytical basis; yet the aggregated effect of instrumental failures, reaction chamber characteristics and waste stream particulars can be illustrated and representatively quantified by empirical means.

7.2 Remote and decentralized operation

In several instances the need for decentralized treatment has been indicated above, such as in the case of stormwater overflow or industrial discharges or the like. Presently the concept of wastewater treatment relies heavily on human supervision and, for reasons of economics, on centralization on the site of the municipal treatment plant.

Automatic control of a treatment process is a necessary prerequisite for decentralization. The problem of manpower then will be of less importance. In particular in the field of treating stormwater overflows the technical and economic needs for decentralized treatment have become evident. This has led to repeated attempts to install such self-controlling units at stormwater discharge points. The treatment units were in all instances similarly conceived and consisted of chemical dosing plus liquid–solid separation.

The very fact that this technology has not yet been applied more broadly testifies that there is still some 'development' needed. It also shows clearly that the most promising line of treatment has again and again been postulated as chemical treatment [15]. Thus, more applied research or development in this area will help to define the limits of application of the chemical treatment concept in this instance.

7.3 Combination of coagulating chemicals with adsorbing chemicals for the removal of dissolved (micro-)pollutants

There are two types of observations that suggest a variation of the classical coagulation flocculation process: first, that the formed aggregates frequently separate very reluctantly from the liquid phase in the sedimentation tank, i.e. their weight is too low, and, second, that there is a tendency to adsorb dissolved material. This has led very early to attempts to dose in addition to dissolved chemicals specific non-dissolved adsorbants which make the floc heavier (such as clays or metal oxides, etc.). Parallel to this development the adsorption process *per se* has been investigated on a technical level in view of the different adsorbant materials. Activated clays as well as specifically prepared metal oxides have proven effective in the removal of cationic as well as anionic inorganic species such as heavy metals and phosphate. It could be expected that very specific dissolved organic substances can be collected by such adsorbents.

The combination of these two processes, i.e. adsorption and aggregation plus separation, is a logical conclusion on the basis of existing experience. Again, there is

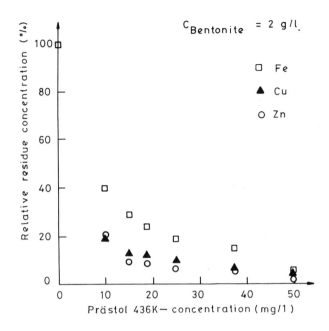
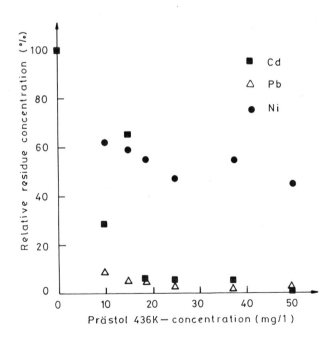

Fig. 12 — Removal of heavy metals through the addition of clay (bentonite) to a coagulation precipitation step (possibly prior to the mechanical-biological treatment).

the problem of combining tested and in most aspects practically proven elements to an overall new concept of treatment with presently unknown long-term behaviour. Pilot plant and large-scale technical investigations are needed.

The result could be a process that is not only effective in removing all suspended material in an economically effective way but also attractive in terms of selectively removing specific dissolved substances such as heavy metals and possibly specific groups of (anthropogenic) higher molecular weight organic material i.e., micropollutants. Figure 12 is a first illustration of the possibilities for the removal of heavy metals by coagulation processes when adsorbents are added [16].

REFERENCES

[1] Bundesministerium des Inneren (Federal Ministry of the Interior of the FRG). *Wirtschaftliche Aspekte der Flockung bei der Abwasserreiningung*, Abschlußbericht zum Vorhaben BMI Wasser 7/75, Bonn, 1975.

[2] Eckenfelder, W. W. Jr, *Wastewater Treatment Design — Part II*, Water and Sewage Works, Vol. 122, pp. 70–92, 1975.

[3] Imhoff, K. R. *Taschenbuch der Stadtenwässerung*, Oldenbourg Verlag, München, 1979.

[4] Abwassertechnische Vereinigung (ATV), Arbeitsblatt A 131, Bonn — St Augustine, 1982.

[5] Bundesminsterium des Inneren, *Erste Schmutwasserverwaltungsvorschrift*, Bonn, 1981.

[6] Umweltbundesamt (Federal Environmental Office). *Möglichkeiten des Entwurfes und des Betriebes von Kläranlagen im Hinblick auf eine Minimierung der Abwasserabgabe' Abschlußberticth zum Forschungsvorhaben Wasser* 31/79, Berlin, 1979.

[7] Dickgiesser, G. J. *Betriebsichere und wirtschaftliche Klärschlammentsorgung*, Schriftenreihe des Institutes für Siedlungswasserwirtschaft, Karlsruhe, Vol. 30, 1982.

[8] Hahn, H. H. *Abwasserfällung/-flockung und Abwasserflotation* in Fortbildungskurs der Abwassertechnischen Vereinigung D3, St. Augustin, 1984.

[9] Optiz, R. *Studie über den Schlammanfall und die Schlammbehandlung bei der Fällungs- und Flockungsreinigung in der BRD* Vertieferarbeit Fakultät für Bauingenieur- und Vermessungswesen der Universität Karlsruhe, Karlsruhe, 1984.

[10] Bischofsberger, W., Ruf, M., Overath, H. and Hegemann, W. *Anwendung von fällungsverfahren zur Verbesserung der Leistungsfähigkeit biologischer anlagen*, Berichte aus Wasserütewirtschaft und Gesundheitsingenieurwesen der TU München, Vol. 13. 1976.

[11] Ives, K.J. (ed.) *The Scientific Basis of Flocculation*, Martinus Nijhoff Publishers, The Hague, 1982.

[12] Culp, R. L. and Culp, G. L. *Advanced Wastewater Treatment*, Van Nostrand-Reinhold Environmental Engineering Series, New York, 1971.

[13] Weber, J. W. Jr, *Physicochemical Processes for Water Quality Control*, Wiley-Interscience, New York, 1972.

[14] Brodkey, K. R. *Turbulence in Mixing Operations*, Academic Press, New York, 1975.
[15] Hansen, C. A. and Agnew, R. W. *Two Wisconsin cities treat combined sewer overflows*, Water and Sewage Works, August 1973, pp. 48–85, 1973.
[16] Xanthopoulos, J. *Schwermetallentfernung mit kombiniertem Flokkungs- und Absorptionsverfahren*, Vertieferarbeit an der Fakultät für Bauingenieur- und Vermessungswesen der Universität Karlsruhe, Karlsruhe, 1985.

4.3

Wastewater treatment by fixed films in biological aerated filters

J. Sibony
OTV, 11 Avenue Dubonnet, 92401 Courbevoie, France

1. INTRODUCTION

Conventional urban and industrial sewage treatment plants are not without a number of weak links that must be improved in order to enhance overall reliability. Our experience in the operation and construction of wastewater treatment plants confirms the opinion of engineers in the profession as a whole that these weak links are generally situated at the following levels:

(1) pre-treatment;
(2) reliability of biological treatment efficiency;
(3) effectiveness of secondary clarification (e.g. bulking problems);
(4) odour control; and
(5) sludge treatment.

Although it does not bring solutions to all such problems, the replacement of activated-sludge treatment by modern fixed biomass systems does undoubtedly help to solve some of them (points 2 and 4) and actually eliminates others (point 3), if properly implemented.

The purpose of this chapter is to describe our particular process, known in France as the Biocarbone filter and in the United States as the biological aerated filter (BAF) and to present some results obtained from a few industrial applications of this process.

2. OUTLINE OF OPERATING PRINCIPLE

The Biocarbone process (see Fig. 1) consists of a downflow filter with a media of sufficiently small grain size to obtain efficient filtration. This media also serves as a

Ch. 4.3] Wastewater treatment by fixed films in biological aerated filters 171

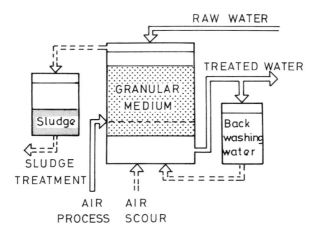

Fig. 1 — Biocarbone flow sheet.

support for the bacteria. The oxygen required to ensure aerobic biological activity is supplied by blowing air into the media through an air manifold located 20–30 cm above the bottom of the filter. This arrangement ensures a good distribution of the air throughout the filter, and it leads to a high transfer efficiency as a result of the tortuous passages, impacts and the accumulation of bubbles on the grains of media. The system depends on a submerged bed to guarantee the hydrostatic conditions necessary for filtration.

The aeration process, by constantly agitating most of the media, allows the suspended solids contained in the influent to penetrate, thus increasing the storage capacity between backwashings. The process can be applied to water issuing from primary sedimentation and sometimes even to raw influent. The layer of media underneath the aerated level has a polishing effect on the suspended solids.

As the headloss in the filter gradually increases, owing to the growth of biomass and the capture of suspended solids, it must be regularly backwashed, the normal cycle time being 24–48 h. The backwashing technique is comparable with that of rapid sand filters, using both air and water. A complete backwash cycle lasts 20 min and leaves a sufficient amount of biomass fixed to the surface of the grains of media to immediately obtain treated water of an acceptable quality.

This process can be applied to the secondary treatment of urban effluents for the removal of BOD, COD and SS and to simultaneous secondary and tertiary treatments when nitrification, with or without denitrification, is added to the above treatments. It can also be used after conventional biological treatment as a nitrifying method or in the treatment of industrial effluents. The same filter is used in all cases, the only changes being the loading rate and aeration parameter.

3. TYPICAL RESULTS ON URBAN SEWAGE (see Figs 2 and 3)

Air required: 25–30 m^3 air/kg COD removed (without nitrification) (=60 m^3/kg BOD)

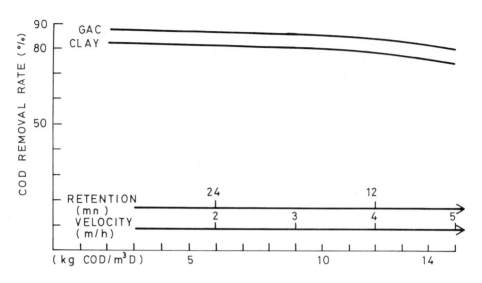

Fig. 2 — COD removal effectiveness as a function of the load.

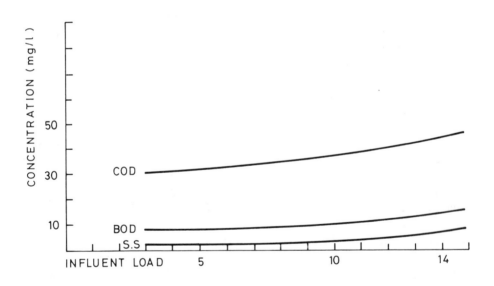

Fig. 3 — Effluent quality as a function of influent load.

Sludge generated: 0.3–0.35 kg TSS/kg BOD removed (=0.85 kg/kg BOD)
Energy consumption (including one backwash per day): 0.14 kWh/m³ treated water
0.4 kWh/kg COD removed (0.9 kWh/kg BOD) for secondary treatment without

carbonaceous removal.
0.28 kWh/m^3 treated water with nitrifying secondary treatment.
0.14 kWh/m^3 treated water in tertiary nitrification treatment.

4. ADVANTAGES TO BE ANTICIPATED

The most obvious advantages to be obtained from such a process result from the principle itself. These were demonstrated during the initial pilot studies (1976 to 1980) and confirmed on the industrial applications subsequently implemented.

4.1 Exceptional treatment quality

With regard to capture of suspended solids, the effluent contains, as a rule, less than 10 mg/litre. The quality of treatment obtained in a single stage is comparable with that of conventional treatment plus filtration. In addition to the attractive visual aspects entailed, this low SS content enables a reduction of about 30% in the quantity of chlorine and ozone necessary for efficient disinfection, compared with that which is needed after treatment with activated sludge.

Secondary settling tanks, the source of many problems in conventional activated-sludge type wastewater treatment plants are no longer required.

4.2 Space saving

Biocarbone is a compact process. Owing to its higher loading rate it requires structures 5–10 times smaller than the conventional biological treatment plant (activated sludge or trickling filters). This is particularly attractive element in cases where:

— the plants must be roofed to avoid odour or because of cold weather conditions;
— available land is limited; and
— the soil has poor mechanical resistance, requiring costly civil works (piles, etc.).

4.3 Rapid achievement of performance ratings

Fixed biomass achieves design performance in one week for carbon removal and two weeks for nitrogen removal. The Biocarbone filter can quickly be put back into service after a toxic impact. The ability to reach normal performance levels very quickly is a valuable asset in facilities where seasonal variations are strong.

5. MAJOR OPERATING EXPERIENCES

Tables 1 and 2 are a summary of some of our experience with the Biocarbone process in municipal wastewater treatment.

For nitrification treatment on secondary effluents, the Paris municipal authority is studying the performance of a facility containing four filters and treating 1500 m^3/d with a view to setting up a system of tertiary treatment at Achéres (2 100 000 m/d). We shall soon obtain operating results concerning installations, built in the United States and Japan by our licensees, as well as pilot units in Canada, Australia and South Africa.

Although the experience acquired through the seven full-scale facilities currently

Table 1 — Secondary treatment

	Population equivalent	Raw water characteristics			Treated water characterists		
		BOD	COD	MES	BOD	COD	SS
Grasse	50 000		1082	124		157	23
Hochfelden	35 000	215	450	143	20	80	23
Le Touquet	8 000 winter 53 000 summer	253	480	180	11	48	9
Decazeville	25 000	Under construction					

Table 2 — Secondary treatment and simulations tertiary treatment

	Population equivalent	Raw water characteristics				Treated water characteristics				
		BOD	COD	MES	NH_4	BOD	COD	MES	NH_4	NO_3
Soissons	40 000	317	600	267	30	9	42	6	9	
Luneville	33 000				Under construction					
Valbonne	25 000	170	396	170	30	10	35	5	3.6	4.5

in operation has revealed a few difficulties, to which we have found solutions, it also brought out at least two outstanding advantages that we had not foreseen.

At the Hochfelden works we are able to demonstrate the possibility of eliminating all specific treatment with regard to odour control by sending the air collected from the units responsible for the odour directly to the bicarbonate filter. This air not only helps supply the oxygen required for the biological treatment of the effluent, but at the same time, the volatile organics that are responsible for odour problems are dissolved in the wastewater and degraded by the biomass in the filter.

For regulating the oxygenation process we have developed a device that continuously measures, by means of u.v. absorption, the pollutant load entering the unit. The very short residence time of wastewater in the filter makes it possible to adjust the amount of air needed with great precision. Since the transfer efficiency is already high due to the passage of bubbles between the grains, we can realize a saving in energy of 15–20%.

Mention has already been made of the fact that a few difficulties were encountered in isolated cases. Among about twenty effluents tested on pilot units and approximately ten industrial facilities, we had one case of clogging in the aeration system, due to a series of operating circumstances above and beyond the already problematic scale-forming nature of the influent. As a result we have modified certain details in the process technology to preclude the recurrence of this phenomenon, regardless of the influent quality. To conclude, we feel justified in saying that the application of fixed biomass has found, in the Bicarbone process, a very reliable system. Despite some minor problems experienced in the early applications, the resulting modifications and 'fine tuning' have produced a process which can bring a truly relevant response to a number of problems arising in wastewater treatment.

Part 5
Hazardous/toxic waste

5.1

Technologies for treatment and management of selected hazardous/toxic wastes

L. W. Canter
Environmental and Ground Water Institute, University of Oklahoma, 200 Felgar Street, Norman, Oklahoma 73019, USA.

1. INTRODUCTION

There is increasing information about ground- and surface-water pollution resulting from improper land disposal of hazardous wastes. Many current pollution situations are the result of previous poor practices, or lack thereof, used for hazardous waste landfill site selection. The purpose of this chapter is to summarize some technologies for treatment and management of hazardous wastes that will reduce the quantity and/or change the mobility characteristics of wastes subjected to disposal, beginning with some background information and followed by major sections on waste reduction technologies and waste treatment and disposal technologies. The chapter is largely drawn from an Office of Technology Assessment report dealing with technologies and management strategies for hazardous waste control [1].

The key term to this discussion is hazardous waste. Section 3001 of the Resource Conservation and Recovery Act of 1976 defines hazardous wastes as:

> a solid waste or combination of solid wastes, which because of its quantity, concentration, or physical, chemical, or infectious characteristics may:
> (a) Cause, or significantly contribute to, an increase in mortality or an increase in serious irreversible, or incapacitating reversible illness, or;
> (b) Pose a substantial present or potential hazard to human health or the environment when improperly treated, stored, transported, or disposed of, or otherwise managed.

Hazardous wastes are of concern since they may be lethal, non-degradable, persistent in nature, and/or biologically magnified. One recent estimate has placed the number of United States sites containing hazardous wastes at between 40 000 and

253 000. It has also been estimated that 14 000–90 000 land disposal sites are contaminating aquifers [2].

Classification of wastes may be developed using one or more of several available techniques. The overall objective is to identify the potential impact a particular waste may have when placed in a landfill or otherwise released to the environment. Factors that must be considered include [3].

(a) Waste composition.
(b) Waste characterization with respect to human effects (e.g. toxic, carcinogenic, irritant, etc.).
(c) Waste persistence and degree of stability, based on potential for biological and chemical reactions within the landfill.
(d) Leachability of the waste, and leachate characteristics.
(e) Degree of attenuation of the leachate within the soil-water system.
(f) Waste handling characteristics (e.g. flammability, reactivity, explosiveness, etc.)

2. WASTE REDUCTION TECHNOLOGIES

Technologies directed toward reducing the volume of hazardous wastes give recognition to the fact that, where technically and economically feasible, it is better to reduce the generation of waste than to incur the costs and risks of managing hazardous waste. Waste reduction technologies include segregation of waste components, process modifications, end-product substitutions, and recycling or recovery operations. Many waste reduction technologies are closely linked to manufacturing and involve proprietary information [1]. Table 1 provides a comparison of the four basic waste reduction technologies.

2.1 Source segregation

Source segregation is the simplest and probably the least costly method of reduction. This approach prevents contamination of large volumes of non-hazardous waste by removal of hazardous constituents to form a concentrated hazardous waste. For example, metal-finishing rinse water is rendered non-hazardous by separation of toxic metals. The water can then be disposed of through municipal/industrial sewage systems.

2.2 Process modification

Process modifications are, in general, made on a continuous basis in existing plants to increase production efficiencies, to make product improvements, and to reduce manufacturing costs. These modifications may be relatively small changes in operational methods, such as a change in temperature, in pressure, or in raw material composition, or may involve major changes such as the use of new processes or new equipment. Although process modifications have reduced hazardous wastes, the reduction may not have been the primary goal of the modifications.

To serve as an example, several process options are available for handling waste from the production of vinyl chloride monomers (VCMs). Five alternatives are

Treatment and management of selected hazardous/toxic wastes

Table 1 — A comparison of the four reduction methods [1]

Advantages	Disadvantages
Source segregation or separation	
(1) Easy to implement; usually low investment	(1) Still have some waste to manage
(2) Short-term solution	
Process modification	
(1) Potentially reduce both hazard and volume	(1) Requires R&D effort; capital investment
(2) Moderate-term solution	(2) Usually does not have industry-wide impact
(3) Potential savings in production costs	
End-product substitution	
(1) Potentially industrywide impact — large volume, hazard reduction	(1) Relatively long-term solutions
	(2) Many sectors affected
	(3) Usually a side benefit of product improvement
	(4) May require change in consumer habits
	(5) Major investments required — need growing market
Recovery/recycling	
• *In-plant*	
(1) Moderate-term solution	(1) May require capital investment
(2) Potential savings in manufacturing costs	(2) May not have wide impact
(3) Reduced liability compared to commercial recovery or waste exchange	
• *Commercial recovery (offsite)*	
(1) No capital investment required for generator	(1) Liability not transferred to operator
(2) Economy of scale for small waste generators	(2) If privately owned, must make profit and return investment
	(3) Requires permitting
	(4) Some history of poor management
	(5) Must establish long-term sources of waste and markets
	(6) Requires uniformity in comparison
• *Waste exchange*	
(1) Transportation costs only	(1) Liability not transferred
	(2) Required uniformity in composition of waste
	(3) Requires long-term relationships — two-party involvement

illustrated in Table 2. All five have been demonstrated on a commercial scale. In most cases, the incineration options (either recycling or add-on treatment) would be selected over chlorinolysis and catalytic fluidized bed reactors.

2.3 End-product substitution

End-product substitution is the replacement of hazardous waste-intensive products (i.e. industrial products, the manufacture of which involves significant hazardous waste) by a new product, the manufacture of which would eliminate or reduce the generation of hazardous waste. Such waste may arise from the ultimate disposal of the product (e.g. asbestos products) or during the manufacturing process (e.g. cadmium plating). Table 3 illustrates six examples of end-product substitution, each

Table 2 — Advantages and disadvantages of process options for reduction of waste streams for VCM manufacture [1]

Treatment option	Type	Advantages	Disadvantages
High-efficiency incineration of vent gas only	Add-on treatment	(1) Relatively simple operation (2) Relatively low capital investment	(1) Second process required to handle liquid waste stream
High-efficiency incineration without HCl recovery	Add-on treatment	(1) Relatively simple operation (2) Relatively low capital	(1) Loss of HCl
High-efficiency incineration with HCl recovery	Recycling	(1) Heat recovery (2) Recover both gaseous and liquid components (3) High reliability	(1) Exit gas requires scrubbing (2) Requires thorough operator training (3) Auxilliary fuel requirements
Chlorinolysis	Modification of process	(1) Carbon tetrachloride generated	(1) High temperatures and pressures required (2) High capital investment costs (3) Weakening market for carbon tetrachloride
Catalytic fluidized bed reactor	Recycling	(1) Low temperature (2) Direct recycle of exit gas (no treatment required)	(1) Limited to oxychlorination plants

representing a different type of problem. General problems include: not all of the available substitutes avoid the production of hazardous waste, and substitutions may not be possible in all situations.

2.4 Recovery and recycling

Recovery of hazardous materials from process effluent followed by recycling provides an excellent method for reducing the volume of hazardous waste. These are not new industrial practices. Recovery and recycling are often used together, but technically the terms are different. Recovery involves the separation of a substance from a mixture. Recycling is the use of such a material recovered from a process effluent [1]. Recovery and recycling operations can be divided into three categories:

(1) In-plant recycling is performed by the waste (or potential waste) generator, and is defined as recovery and recycling of raw materials, process streams or by-products for the purpose of prevention or elimination of hazardous waste. If several products are produced at one plant by various processes, materials from the effluent of one process may become raw materials for another through in-plant recycling. An example is the recovery of relatively dilute sulphuric acid, which is then used to neutralize an alkaline waste. In-plant recycling offers

Table 3 — End-product substitutes for reduction of hazardous waste [1]

Product	Use	Ratio of waste:[a] original product	Available substitute	Ratio of waste:[a] substitute product
Asbestos	Pipe	1.09	Iron Clay PVC	0.1 phenol, cyanides 0.05 fluorides 0.04 VCM manufacture+ 1.0 PVC pipe
	Friction products (brake linings)	1.0+manufacturing waste	Glass fibre Steel wool Mineral wools Carbon fibre Sintered metals Cement	0
	Insulation	1.0+ manufacturing	Glass fibre Cellulose fibre	0.2
PCBs	Electrical transformers	1.0	Oil-filled transformers Open-air-cooled transformers	0 0
Cadmium	Electroplating	0.29	Zinc electroplating	0.06
Creosote treated wood	Piling		Concrete, steel	0.0 (reduced hazard)
Chlorofluorocarbons	Industrial solvents	70/81 = 0.9	Methyl chloroform; methylene chloride	0.9 (reduced hazard)
DDT	Pesticide	1.0+manufacturing waste	Other chemical pesticides	(reduced hazard) 1.0+manufacturing waste

[a] Quantity of hazardous waste generated/unit of product.

several benefits to the manufacturer, including savings in raw materials, energy requirements, and disposal or treatment costs.

(2) Commercial (offsite) recovery can be used for those wastes combined from several processes or produced in relatively small quantities by several manufacturers. Commercial recovery means that an agent other than the generator of the waste is handling collection and recovery.

(3) Material exchanges (often referred to as 'waste' exchanges) are a means to allow material users to identify waste generators producing a material that could be used. Waste exchanges are listing mechanisms only and do not include collection, handling, or processing. Although benefits occur by elimination of disposal and treatment costs for a waste as well as receipt of cash value for a waste, responsibility for meeting purchaser specifications remains with the generator.

A number of existing technologies are useful for recovery and/or recycling. These technologies can be divided into physical separation, component separation, and chemical transformation methods (Table 4).

Table 4 — Description of technologies currently used for recovery of materials [1]

Technology/ description	Stage of development	Economics	Types of waste streams	Separation efficiency[a]	Industrial applications
Physical separation:					
Gravity settling: Tanks, ponds provide hold-up time allowing solids to settle; grease skimmed to overflow to another vessel	Commonly used in wastewater treatment	Relatively inexpensive; dependent on particle size and settling rate	Slurries with separate phase solids, such as metal hydroxide	Limited to solids (large particles) that settle quickly (less than 2 hours)	Industrial wastewater treatment first step
Filtration: Collection devices such as screens, cloth, or other; liquid passes and solids are retained on porous media	Commonly used	Labour intensive: relatively inexpensive; energy required for pumping	Aqueous solutions with finely divided solids; gelatinous sludge	Good for relatively large particles	Tannery water
Flotation: Air bubbled through liquid to collect finely divided solids that rise to the surface with the bubbles	Commercial application	Relatively inexpensive	Aqueous solutions with finely divided solids	Good for finely divided solids	Refinery (oil/water mixtures); paper waste; mineral industry
Flocculation: Agent added to aggregate solids together which are easily settled	Commerical practice	Relatively inexpensive	Aqueous solutions with finely divided solids	Good for finely divided solids	Refinery; paper waste; mine industry
Centrifugation: Spinning of liquids and centrifugal force causes separation by different densities	Practised commercially for small-scale systems	Competitive with filtration	Liquid/liquid or liquid/solid separation, i.e., oil/water; resins; pigments from lacquers	Fairly high (90%)	Paints
Component separation:					
Distillation: Successfully boiling off of materials at different temperatures (based on	Commerical practice	Energy intensive	Organic liquids	Very high separations achievable (99+% concentrations) of several components	Solvent separations; chemical and petroleum industry

Ch. 5.1] Treatment and management of selected hazardous/toxic wastes

Method	Status	Cost/Energy	Waste type	Effectiveness	Applications
Evaporation: Solvent recovery by boiling off the solvent	Commercial practice in many industries	Energy intensive	Organic/inorganic aqueous streams; slurries, sludges, i.e., caustic soda	Very high separations of single, evaporated component achievable	Rinse waters from metal plating waste
Ion exchange: Waste stream passed through resin bed, ionic materials selectively removed by resins similar to resin adsorption. Ionic exchange materials must be regenerated	Not common for HW	Relatively high costs	Heavy metals aqueous solutions; cyanide removed	Fairly high	Metal-plating solutions
Ultrafiltration: Separation of molecules by size using membrane	Some commercial application	Relatively high	Heavy metal aqueous soltions	Fairly high	Metal-coating applications
Reverse osmosis: Separation of dissolved materials from liquid through a membrane	Not common: growing number of applications as secondary treatment process such as metal-plating pharmaceuticals	Relatively high	Heavy metals; organics; inorganic aqueous solutions	Good for concentrations less than 300 p.p.m.	Not used industrially
Eltrolysis: Separation of positively/negatively charged materials by application of electric current	Commercial technology; not applied to recovery of hazardous materials	Dependent on concentrations	Heavy metals; ions from aqueous solutions; copper recovery	Good	Metal plating
Carbon/resin adsorption: Dissolved materials selectively adsorbed in carbon or resins. Adsorbents must be regenerated	Proven for thermal regeneration of carbon; less practical for recovery of adsorbate	Relatively costly thermal regeneration; energy intensive	Organics/inorganics from aqueous solutions with low concentrations, i.e. phenols	Good, overall effectiveness dependent on regeneration method	Phenolics

Table 4 — (contd.)

Technology/description	Stage of development	Economics	Types of waste streams	Separation efficiency[a]	Industrial applications
Solvent extraction: Solvent used to selectively dissolve solid or extract liquid from waste	Commonly used in industrial processing	Relatively high costs	Organic liquids, phenols, acids	Fairly high loss of solvent may contribute to hazardous waste problem	Recovery of dyes
Chemical transformations:					
Precipitation: Chemical reaction causes formation of solids which settle	Common	Relatively high costs	Lime slurries	Good	Metal-plating wastewater treatment
Electrodialysis: Separation based on differential rates of diffusion through membranes. Electrical current applied to enhance ionic movement	Commercial technology, not commercial for hazardous material recovery	Moderately expensive	Separation/concentration of ions from aqueous streams; application to chromium recovery	Fairly high	Separation of acids and metallic solutions
Chlorinolysis: Pyrolysis in atmosphere of excess chlorine	Commerically used in West Germany	Insufficient US market for carbon tetrachloride	Chlorocarbon waste	Good	Carbon tetrachloride manufacturing
Reduction: Oxidative state of chemical changed through chemical reaction	Commerically applied to chromium; may need additional treatment	Inexpensive	Metals, mercury in dilute streams	Good	Chrome-plating solutions and tanning operations
Chemical dechlorination: Reagents selectively attack carbon–chlorine bonds	Common	Moderately expensive	PCB-contaminated oils	High	Transformer oils
Thermal oxidation: Thermal conversion of components	Extensively practised	Relatively high	Chlorinated organic liquids, silver	Fairly high	Recovery of sulphur, HCl

[a] Good implies 50 to 80% efficiency, fairly high implies 80%, and very high implies 90%.

3. WASTE TREATMENT AND DISPOSAL TECHNOLOGIES

Waste treatment and disposal technologies are useful for reducing the hazard of the waste. These two groupings of technologies contrast distinctly, in that it is preferable to permanently reduce risks to human health and the environment by waste treatments that destroy or permanently reduce the hazardous character of the material than to rely on long-term containment in land-based disposal structures [1]. Table 5 provides a summary comparison of some hazard reduction technologies in accordance with five generic groupings.

3.1 Thermal treatment

There are a variety of treatment technologies involving high temperatures which have, or are likely to have, important roles in hazardous waste management. Most of these technologies involve combustion, but some are more accurately described as destruction by i.r. or u.v. radiation. Table 6 has summary comparisons of thermal treatment technologies for hazard reductions.

3.2 Chemical stabilization

The objective of solidification/stabilization processes is to fix chemically the waste in a solid matrix. This reduces the exposed surface area and minimizes leaching of toxic constituents. Effective immobilization includes reacting toxic components chemically to form compounds immobile in the environment and/or entrapping the toxic material in an inert stable solid. From a definitional perspective, stabilization refers to immobilization by chemical reaction or entrapping (watertight inert polymer or crystal lattice), while solidification means the production of a solid, monolithic mass with sufficient integrity to be easily transported.

Probably the simplest stabilization process is pH adjustment. In most industrial sludges, toxic metals are precipitated as amorphous hydroxides that are insoluble at an elevated pH. By carefully selecting a stabilization system of suitable pH, the solubility of any metal hydroxide can be minimized. Certain metals can also be stabilized by forming insoluble carbonates or sulphides. Care should be taken to ensure that these metals are not remobilized because of changes in pH or redox conditions after they have been introduced into the environment. Stabilized wastes can be solidified into a solid mass by micro- or macro-encapsulation. Microencapsulation refers to the dispersion and chemical reaction of the toxic materials within a solid matrix. Therefore, any breakdown of the solid material only exposes the material located at the surface to potential release to the environment. Macroencapsulation is the sealing of the waste in a thick, relatively impermeable coating layer. Plastic and asphalt coatings, or secured landfilling, are considered to be macroencapsulation methods. Breakdown of the protective layer with macroencapsulation could result in a significant release of toxic material to the environment. Present solidification/stabilization systems can be grouped into seven classes or processes:

(1) Solidification through cement addition;
(2) Solidification through the addition of lime and other pozzolanic materials;

Table 5 — Comparison of some hazard reduction technologies [1]

	Disposal		Treatment		
	Landfills and impoundments	Injection wells	Incineration and other thermal destruction	Emerging high-temperature decomposition	Chemical stabilization
Effectiveness: How well it contains or destroys hazardous characteristics	Low for volatiles, question for liquids; based on lab and field tests	High, based in theory, but limited field data available	High, based on field tests, except little data on specific constituents	Very high, commercial-scale tests	High for many metals, based on lab tests
Reliability issues:	Siting, construction, and operation Uncertainties: long-term integrity of cells and cover, liner life less than life of toxic waste	Site history and geology; well depth, construction and operation	Long experience with design Monitoring uncertainties with respect to high degree of DRE; surrogate measures, PICs, incinerability	Limited experience Mobile units; onsite treatment avoids hauling risks Operational simplicity	Some inorganics still soluble Uncertain leachate test, surrogate for weathering
Environmental media most affected:	Surface and ground water	Surface and ground water	Air	Air	None likely
Least compatible waste[b]	Linear reactive; highly toxic, mobile persistent, and bioaccumulative	Reactive; corrosive; highly toxic, mobile, and persistent	Highly toxic and refractory organics, high heavy metals concentration	Possibly none	Organics
Costs: Low, Mod. High	L–M	L	M–H (Coincin. = L)	M–H	M
Resource recovery: Potential	None	None	Energy and some acids	Energy and some metals	Possible building material

[a] Molten salt, high-temperature fluid wall, and plasma arc treatments.
[b] Waste for which this method may be less effective for reducing exposure, relative to other technologies. Waste listed does not necessarily denote common usage.

Ch. 5.1] Treatment and management of selected hazardous/toxic wastes

Currently available incinerator designs:

Liquid injection incineration:
Can be designed to burn a wide range of pumpable waste. Often used in conjunction with other incinerator systems as a secondary afterburner for combustion of volatilized constituents. Hot refractory minimizes cool boundary layer at walls. HCl recovery possible.

Limited to destruction of pumpable waste (viscosity of less than 10000 SSI). Usually designed to burn specific waste streams. Smaller units sometimes have problems with clogging of injection nozzle.

Estimated that 219 liquid injection incinerators are in service, making this the most widely used incinerator design.

Rotary kilns:
Can accommodate great variety of waste feeds: solids, sludges, liquids, some bulk waste contained in fibre drums. Rotation of combustion chamber enhances mixing of waste by exposing fresh surfaces for oxidation.

Rotary kilns are expensive. Economy of scale means regional locations, thus, waste must be hauled increasing spill risks.

Estimated that 42 rotary kilns are in service under interim status. Rotary kiln design is often centre-piece of integrated commercial treatment facilities. First non-interim RCRA permit for a rotary kiln incinerator (IT Corp.) is currently under review.

Cement kilns:
Attractive for destruction of harder-to-burn waste, due to very high residence times, good mixing, and high temperatures. Alkaline environmental neutralizes chlorine.

Burning of chlorinated waste limited by operating requirements, and appears to increase particulate generation. Could require retrofitting of pollution control equipment and of instrumentation for monitoring to bring existing facilities to comparable level. Ash may be hazardous residual.

Cement kilns are currently in use for waste destruction, but exact number is unknown. National kiln capacity is estimated at 41.5 million tonnes/yr. Currently mostly non-halogenated solvents are burned.

Boilers (usually a liquid injection design):
Energy value recovery, fuel conservation. Availability on sites of waste generators reduces spill risks during hauling.

Cool gas layer at walls result from heat removal. This contains design to high efficiency combustion within the flame zone. Nozzle maintenance and waste feed stability can be critical. Where HCl is recovered, high temperatures must be avoided. (High temperatures are good for DRE.) Metal parts corrode where halogenated waste are burned.

Boilers are currently used for waste disposal. Number of boiler facilities is unknown, quantity of wastes combusted has been roughly estimated at between 17.3 to 20 million tonnes/yr.

Applications of currently available designs:

Multiple hearth:
Passage of waste onto progressively hotter hearths can provide for long residence times for sludges. Design provides good fuel efficiency. Able to handle wide variety of sludges.

Tiered hearths usually have some relatively cold spots which inhibit even and complete combustion. Opportunity for some gas to short circuit and escape without adequate residence time. Not suitable for waste streams which produce fusible ash when combusted. Units have high maintenance requirements due to moving parts in high-temperature zone.

Technology is available; widely used for coal and municipal waste combustion.

Table 6 — *(contd.)*

Advantages of design features	Disadvantage of design features	Status for hazardous waste treatment
Fluidized-bed incinerators: Turbulence of bed enhances uniform heat transfer and combustion of waste. Mass of bed is large relative to the mass of injected waste.	Limited capacity in service. Large economy of scale.	Estimated that nine fluidized-bed incinerators are in service. Catalytic bed may be developed.
At-sea incineration: shipboard (usually liquid injection incinerator): Minimum scrubbing of exhaust gases required by regulations on asssumption that ocean water provides sufficient neutralization and dilution. This could provide economic advantages over land-based incineration methods. Also, incineration occurs away from human populations. Shipboard incinerators have greater combustion rates; e.g., 10 tonnes/h.	Not suitable for wastes that are shock sensitive, capable of spontaneous combustion, or chemically or thermally unstable, due to the extra handling and hazard of shipboard environment. Potential for accidental release of waste held in storage (capacities vary from between 4000 to 8000 tonnes).	Limited burns of organochlorine and PCB were conducted at sea in mid-1970. PCB test burns conducted by Chemical Waste Management, Inc., in January 1982 are under review by EPA. New ships under construction by At Sea incineration, inc.
At-sea incineration: oil drilling platform-based: Same as above, except relative stability of platform reduces some of the complexity in designing to accommodate rolling motion of the ship.	Requires development of storage facilities. Potential for accidental release of waste held in storage.	Proposal incinerator currently under review by EPA.
Pyrolysis: Air pollution control need minimum: air-starved combustion avoids volatilization of any inorganic compounds. These and heavy metals go into insoluble solid char. Potentially high capacity.	Greater potential for PIC formation. For some waste produce a tar which is hard to dispose of. Potentially high fuel maintenance cost. Waste-specific designs only.	Commercially available but in limited use.
Emerging thermal treatment technologies: Molten salt: Molten salts act as catalysts and efficient heat transfer medium. Self-sustaining for some wastes. Reduces energy use and reduces maintenance costs. Units are compact; potentially portable. Minimal air pollution control needs; some combustion products, e.g., ash and acidic gases are retained in the melt.	Commercial-scale applications face potential problems with regeneration or disposal of ash-contaminated salt. Not suitable for high ash wastes. Chamber corrosion can be a problem. Avoiding reaction vessel corrosion may imply tradeoff with DRE.	Technology has been successful at pilot plant scale, and is commercially available.

Technology	Description	Limitations	Status
High-temperature fluid wall:	Waste is efficiently destroyed as it passes through cylinder and is exposed to radiant heat temperatures of about 4000°F. Cylinder is electrically heated; heat is transferred to waste through inert gas blanket, which protects cylinder wall. Mobile units possible.	To date, core diameters (3 in, 6 in, and 12 in) and cylinder length (72 in) limit throughput capacity. Scale-up may be difficult due to thermal stress on core. Potentially high costs for electrical heating.	Other applications tested; e.g., coal gasification, pyrolysis of metal-bearing refuse and hexachlorobenzene. Test burns on toxic gases in December 1982.
Plasma arc:	Very high energy radiation (at 50000°F) breaks chemical bonds directly, without series of chemical reactions. Extreme DREs possible, with no or little chance of PICs. Simple operation, very low energy costs, mobile units planned.	Limited throughput. High use of NaOH for scrubbers.	Limited US testing, but commercialization in July 1983 expected. No scale-up needed.
Wet oxidation:	Applicable to aqueous waste too dilute for incineration and too toxic for biological treatment. Lower temperatures required, and energy released by some wastes can produce self-sustaining reaction. No air emissions.	Not applicable to highly chlorinated organics, and some wastes need further treatment.	Commercially used as pretreatment to biological wastewater treatment plant. Bench-scale studies with catalyst for nonchlorinated organics.
Supercritical water:	Applicable to chlorinated aqueous waste which are too dilute to incinerate. Takes advantage of excellent solvent properties of water above critical point for organic compounds. Injected oxygen decomposes smaller organic molecules to CO_2 and water. No air emissions.	Probable high economy of scale. Energy needs may increase on scale-up.	Bench-scale success (99.99% DRE) for DDT, PCBs, and hexachlorobenzene.

(3) Techniques involving embedding wastes in thermoplastic materials such as bitumen, paraffin, or polyethylene;
(4) Solidification by addition of an organic polymer;
(5) Encapsulation of wastes in an inert coating;
(6) Treatment of the wastes to produce a cementitious product with major additions of other constituents; and
(7) Formation of glass by fusion of wastes with silica.

3.3 Physical, chemical and biological treatment processes

Many physical, chemical, and/or biological treatment processes can be used to eliminate or reduce the hazardous attributes of wastes. Several physical and chemical processes are listed in Table 4; Table 7 lists several biological treatment methods. All

Table 7 — Conventional bilogical treatment methods [1]

Treatment method	Aerobic (A) anaerobic (N)	Waste applications	Limitations
Activated sludge	A	Aliphatics, aromatics, petrochemicals, steelmaking, pulp and paper industries	Volatilization of toxics; sludge disposal and stabilization required
Aerated lagoons	A	Soluble organics, pulp and paper, petrochemicals	Low efficiency due to anaerobic zones, seasonal variations; requires sludge disposal
Trickling filters	A	Suspended solids, soluble organics	Sludge disposal required
Biocontactors	A	Soluble organics	Used as secondary treatment
Packed bed reactors	A	Nitrification and soluble organics	Used as secondary treatment
Stabilization ponds	A&N	Concentrated organic waste	Inefficient; long retention times, not applicable to aromatics; sludge removal and disposal required
Anaerobic digestion	N	Non-aromatic hydrocarbons; high-solids; methane generation	Long retention times required; inefficient on aromatics
Landfarming/spreading	A	Petrochemicals, refinery waste, sludge	Leaching and runoff occur; seasonal fluctutations; requires long retention times
Composting	A	Sludges	Volatilization of gases, leaching, runoff occur; long retention time; disposal of residuals

Aerobic — requires presence of oxygen for cell growth.
Anaerobic — requires absence of oxygen for cell growth.

of these methods produce waste residuals, usually a liquid and a solid waste. The hazardous characteristics of these waste residuals must be evaluated in terms of the objective desired for their final disposition or recovery.

3.4 Landfills

Landfilling is the burial of waste in excavated trenches or cells. The waste may be in bulk form or containerized. In the early 1970s, landfills specifically designed to contain industrial waste were constructed. Experience with the operation and construction of these more advanced landfills has been an evolutionary process, and is ongoing. Over time, fractions of the waste can be released from the landfill, either as leachate or as volatilized gases. The objective of landfilling design is to reduce the frequency of occurrence of releases so that the rate of release does not impair water or air resources. Liquids are able to leak through compacted clays or synthetic lining materials. Reducing the potential for migration of toxic constituents from a landfill requires minimizing the production of liquids and controlling the movement of those that inevitably form [1].

A landfill has three primary engineered control features: a bottom liner(s), a leachate collection system, and a cover [1]. The bottom liner(s) retard the migration of liquids and leachate from the landfill cells. Bottom liners are constructed of compacted clay, a clay and soil mixture, or synthetic material — often synthetic membranes. Leachate is collected through a series of pipes buried in a drainage bed placed above the bottom liner. A mechanical pump raises the leachate through standpipes to the surface. The final cover reduces infiltration of precipitation into the closed landfill. Intermediate covers can be applied for the same purpose during operation of the landfill. Table 8 summarizes function and failure mechanisms for each of these components.

3.5 Surface impoundments

Surface impoundments are depressions in the ground used to store, treat, or dispose of a variety of industrial wastes. They have a variety of names: lagoons, treatment basins, pits and ponds. These depressions can be natural, man-made, lined or unlined. They can be several meters in diameter or hundreds of hectares in size.

Surface impoundments have allowed release of hazardous waste constituents through catastrophic failure, leachate migration, and volatization of organics. Impoundments are more subject to catastrophic failure than landfills because they tend to contain more bulk liquid. Evidence of surface and ground water contamination resulting from impoundments is well documented [1]. This has occurred from sudden releases such as overtopping the sides, dike failures, or rupture of the liner due to inadequate subgrade preparation, or sinkhole formation. In addition, slow leakage can contaminate soil and ground water. This is especially true for unlined impoundments. Investigations at some unlined 'evaporation ponds' have shown that seepage accounted for more of the reduction in volume than did evaporation [1].

There is an expected rate of leakage even through intact liners. Some liquids are chemically aggressive to liners, increasing the rate of movement through the liner. In addition, there is always a hydraulic gradient acting on the liner [1].

Table 8 — Engineered components of landfills: their function and potential causes of failure [1]

Function	Potential causes of failure
Cover: To prevent infiltration of precipitation into landfill cells. The cover is constructed with low permability synthetic and/or clay material and with graded slopes to enhance the diversion of water.	• After maintenance ends, cap integrity can be threatened by desiccation, deep rooted vegetation, animals, and human activity. • Wet/dry and freeze/thaw cycles, causing cracking and increased infiltration. • Erosion; causing exposure of cover material to sunlight, which can cause polymeric liners to shrink, break, or become brittle. • Differential settling of the cover, caused by shifting, settling, or release of the landfill contents over time. Settling can cause cracking or localized depressions in the cover, allowing ponding and increased infiltration.
Leachate collection and recovery system: To reduce hydrostatic pressure on the bottom liner, and reduce the potential for flow or leachate through the liner. Leachate is collected from the bottom of the landfill cells or trenches through a series of connected drainage pipes buried within a permable drainage layer. The collection leachate is raised to the surface by a mechanical pump.	• Clogging of drainage layers or collection pipes. • Crushing of collection pipes due to weight of overlying waste. • Pump failures.
Bottom liner: To reduce the rate of leachate migration to the subsoil.	• Faulty installation, damage during or after installation. • Deformation and creep of the liner on the sloping walls of the landfill. • Differential settling, most likely to where landfill is poorly sited or subgrade is faulty. • Structural failure of the liner in response to hydrostatic pressure. • Degradation of liner material resulting from high strength chemical leachate or microbial action. • Swelling of polymeric liners, resulting in loss of strength and puncture resistance. • Chemical extraction of plasticizers from polymer liners.

3.6 Injection wells

Injection of liquid wastes into sub-surface rock formations is a technology that uses porous sedimentary strata to hold liquid waste. The pores of all porous rock formations contain liquids, gases or both. The gas or liquid is contained within the strata under pressure caused by overlying rocks. Internal pressures within strata can vary significantly, depending on the porosity of the formation, its depth, and other physical and chemical factors. Essentially, underground injection entails drilling a well to the depth required to intersect an appropriate geologic formation (known as the injection zone) and pumping the liquid waste in with pressure sufficient to

Ch. 5.1] Treatment and management of selected hazardous/toxic wastes

displace the native fluids, but not so great as to cause fracturing of the strata or excessive migration of the waste. A formation suitable for waste injection should meet the following criteria [1]:

(1) it should not have value as a resource, e.g. as a source of drinking water, hydrocarbons, or geothermal energy;
(2) it must have sufficient porosity and volume to be able to accept the anticipated amount of liquids;
(3) it should be sealed both above and below by formations with sufficient strength, thickness, and impermeability to prevent migration of the waste from the disposal zone; and
(4) it should be located in an area with little seismic activity to minimize both the risk of earthquake damage to the well and triggering of seismic events.

There is no standard injection-well design because design requirements are influenced by site-specific geology. Figure 1 illustrates the design of an injection well

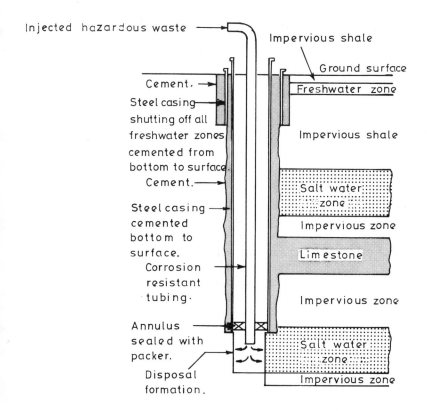

Fig. 1 — Schematic of typical completion method for a deep waste injection well (after ref. [4]).

that might be used for hazardous waste disposal [4]. As shown in Fig. 1, the well is constructed with three concentric casings: the exterior surface casing, the intermediate protection casing, and the injection tubing. The exterior surface casing is designed to protect freshwater in the aquifiers through which the well passes and to protect the well exterior from corrosion. The casing extends below the base of aquifers containing potable water and is cemented along its full length. Similarly, the intermediate protection casing extends down and through the top of the injection zone and is cemented along its full length. The waste is actually transported through the injection tubing, the innermost casing. The tubing also extends into the top of the injection zone; its endpoint is the point of waste discharge. The injection tubing is sealed off from the intermediate casing, creating annular space between the injection tubing and the casing. The annulus is filled with fluid containing corrosion inhibitors to protect the casing and tubing metal. The fluid is pressurized between the sealing at the base of the well and the well head assembly. Since the pressure within the annulus is known, monitoring the pressure during the operation of the well can be a method of checking the integrity of the injection system. When injection operations cease, the well should be plugged. Proper plugging is necessary to maintain the existing pressure in the injection zone, to prevent mixing of fluids from different geologic strata, and to prevent flow of liquids from the pressurized zone to the surface [1].

4. SUMMARY

The following points can be made from this review of technologies for treatment and management of hazardous/toxic wastes [1].

(1) Source segregation is the easiest and most economic method of reducing the volume of hazardous waste. This method of hazardous waste reduction has been implemented in many cases, particularly by large industrial firms. Many opportunities still exist for further application.

(2) Through a desire to reduce manufacturing costs by using more efficient methods, industry has implemented various process modifications. Although a manufacturing process often may be used in several plants, each facility has slightly different operating conditions and designs. Thus, a modification resulting in hazardous waste reduction may not be applicable industrywide. Also, proprietary concerns may inhibit information transfer.

(3) Product substitutes generally have been developed to improve performance. Hazardous waste reduction has been a side-benefit, not a primary objective. In the long term, end-product substitution could reduce or eliminate some hazardous wastes.

(4) With regard to recovery and recycling approaches to waste reduction, if extensive recovery is not required prior to recycling a waste constituent, in-plant operations are relatively easy. Commercial recovery benefits are few for medium-sized generators. No investment is required, but liability remains with the generator. Commercial recovery has certain problems as a profit-making enterprise. The operator is dependent on suppliers' waste as raw material; contamination and consistency in composition of a waste are difficult to control. Waste exchanges are not very popular at present, since generators must assume

all liability in transferring waste. Also, small firms do not generate enough waste to make it attractive for recycling.

(5) Many waste treatment technologies can provide permanent, immediate and very high degrees of hazard reduction. In contrast, the long-term effectiveness of land-based disposal technologies relies on continued maintenance and integrity of engineered structures and proper operation. For wastes which are toxic, mobile, persistent, and bioaccumulative, and which are amenable to treatment, hazard reduction by treatment is generally preferable to land disposal. In general however, costs for land disposal are comparable to, or lower than, unit costs for thermal or chemical treatment.

(6) For waste disposal, advanced landfill designs, surface impoundments, and injection wells are likely to perform better than their earlier counterparts. However, there is insufficient experience with these more advanced designs to predict their performance. Site- and land-specific factors and continued maintenance of final covers and well plugs will be important.

REFERENCES

[1] Office of Technology Assessment, *Technologies and Management Strategies for Hazardous Waste Control*, OTA-M-196, US Congress, Washington, DC, 1983.

[2] US Environmental Protection Agency, *Planning Workshops to Develop Recommendations for a Ground Water Protection Strategy, Sections I, II, and III*, Washington, DC, May 1980.

[3] Corbin, M. General considerations for hazardous waste management facilities, in A. A. Metry (ed.), *The Handbook of Hazardous Waste Management*. Technomic Publishing Company, Westport, CT, 1980, pp. 158–192.

[4] Projasek, R. B. *Toxic and Hazardous Waste Disposal*, Vol. 4. Ann Arbor Science Publishers, Ann Arbor, MI, 1980.

5.2

Hazardous waste: status and future trends of treating technologies

F. J. Colon and J. H. O. Hazewinkel
Grabovsky and Poort, The Hague, Netherlands

1. INTRODUCTION

The treatment, recycling and disposal of all sorts of hazardous waste in an economically and environmentally acceptable manner is a complex problem both nationally and internationally. This is particularly the case with densely populated and highly industrialized countries making extensive use of their soil for agricultural purposes, and dependent on ground and surface waters as a source of water, requiring careful handling of hazardous wastes.

Current and newly emerging technologies, practices, and research efforts are strongly influenced by a number of factors, such as:
— Public awareness exerting an influence on nationally imposed regulations and legislation;
— The state of knowledge in the field of toxicology (dose–effect relationships);
— The scarcity of landfill locations;
— The overall economic situation of any particular country;
— The necessity to save raw materials and energy;
— The availability of clean production processes;
— The availability of proper treatment facilities.

As a result of all these factors, often differing from country to country, a proper definition of hazardous waste is a complex matter.

2. DEFINITION OF HAZARDOUS WASTES

Discussions have been going on for years on both a national and an international scale as to what hazardous wastes really are, and how they can be defined and dealt with in a responsible manner. In general, there is a consensus of opinion on what constitutes the most dangerous types of hazardous waste. As the danger to humanity and the environment decreases, so does the level of agreement. The terminology used differs from country to country. In Belgium for example, the substances involved are referred to as *toxic* wastes, in the Netherlands as *chemical* wastes, in the United States as *hazardous* wastes and in Germany as *special* wastes. This simply demonstrates different approaches to the matter and it is clear that the formulation of a proper definition is indeed a highly complex matter. This is also related to the need for, and enforcement of, (inter)national legislation. The definition of hazardous waste is mainly determined by the potentially hazardous effects of the waste on humanity and the environment (from short- and long-term aspects). Short-term effects include hazards such as acute toxicity by ingestion, inhalation or skin absorption, hazards produced by fire or explosion, and corrosion. Long-term effects include carcinogenicity, mutagenicity, teratogenicity, chronic toxicity, the resistance to biodegradation of toxic substances, ground and/or surface water pollution.

The NATO-CCMS pilot study *Disposal of Hazardous Wastes* concluded that the definition of hazardous waste is a reflection of the tendency to express the nature of the environmental problem to be solved in terms of the problems of economic, social and political conditions of the countries involved. The study also concluded that the environmental problems should be taken into consideration with the methods of disposal applied.

In its directive on toxic and dangerous wastes, the Council of European Communities defines toxic and dangerous wastes as any waste containing or contaminated by the substances listed in the Annex to the Directive and of such a nature, and in such quantities or in such concentration, as to constitute a risk to health or to the environment. A list of these hazardous substances is given in Table 1.

3. SOME FACTORS PERTINENT TO THE GENERATION OF HAZARDOUS WASTE

In modern society, a large number of raw materials and processes are used to manufacture a wide variety of products with the consequent generation of waste matter (see Table 2). For many years a careless attitude prevailed with respect to waste products. During that time those responsible for these processes considered waste to be an inevitable and unimportant part of the production process, added to which there was a general lack of awareness about the potential negative consequences for both humanity and the environment. Waste generation was assessed mainly on the basis of the economic factors involved. These attitudes changed as it became clear that the factors of health, safety and environment also play an important role in the course of events.

What are the reasons for hazardous wastes being produced? The selection of materials, processing, manufacturing and commercial sales some of the products/by-products can result in the following:
— Non-saleable by-products (the costs involved in dealing with waste being lower than the revenue from such products);
— Products not meeting specifications (rejects);
— Materials left over from production and dismantling of plant;
— Used packaging materials; and,
— Materials originating from processing and transport accidents.

Table 1 — List of toxic or dangerous substances and materials as given in the appendix to the EEC Directive 78/319/EEC

1. Arsenic; arsenic compounds
2. Mercury; mercury compounds
3. Cadmium; Cadmium compounds
4. Thallium; thallium compounds
5. Beryllium; beryllium compounds
6. Chrome 6 compounds
7. Lead; lead compounds
8. Antimony; antimony compounds
9. Phenols; phenol compounds
10. Cyanides; cyanide compounds
11. Isocyanates
12. Organic-halogen compounds, excluding inert polymeric materials and other substances referred to in this list or covered by other Directives concerning the disposal of toxic or dangerous waste
13. Chlorinated solvents
14. Organic solvents
15. Biocides and phyto-pharmaceutical substances
16. Tarry materials from refining and tar residues from distilling
17. Pharmaceutical compounds
18. Peroxides, chlorates, perchlorates and azides
19. Ethers
20. Chemical laboratory materials, not identifiable and/or new, whose effects on the environment are not known
21. Asbestos (dust and fibres)
22. Selenium; selenium compounds
23. Tellurium; tellurium compounds
24. Aromatic polycyclic compounds (with carcinogenic effects)
25. Metal carbonyls
26. Soluble copper compounds
27. Acids and/or basic substances used in the surface treatment and finishing of metals.

Another cause of the generation of hazardous wastes is the measures taken to prevent water and air pollution. Sludges accruing from flue-gas purification and wastewater treatment processes can, for example, lead to increasing amounts of waste products (Fig. 1). Thus there is a shift from water and air pollution to wastes that contribute to soil and groundwater pollution.

4. HAZARDOUS WASTE AS A PROBLEM

Waste became a problem as a consequence of an unthinking and careless attitude by industry and government alike, in that more waste was produced than existing waste

Status and future trends of treating technologies

Table 2 — A summary of several aspects relating to existence of certain flows of chemical waste

	>90% organic	>90% inorganic	Mixture	Most important pollution	Most important constituents with a potential value	Quantities per year from all sources			Number of sources				Possibilities of			Present processing method
						<1000 ton	>1000 ton	>50000 ton	<10	<10/>50	<50/>500	>500	prevention	application	processing[1]	
Production waste																
GKW's	X			Org. Cl	Org.	X					X		—	O	+	Inciner./reuse
pickling baths		X		Heavy met.	—			X			X		O	O	+	Detoxification
shredder-wastes			X	Heavy met.	C_xH_y			X	X				O	—	O	Deposition
jarosite		X		Heavy met.	Fe-inc.			X	X				O	—	O	Deposition
Cleaning																
waste waters	X			Org.	Org.			X		X			O	—	O	Discharge/proces.
jet grit		X		Heavy met.	Grit			X		X		X	O	O	—	Deposition
Emission reduction																
calcium sludge		X		Heavy met.	Fe-verb.			X	X				—	O	O	Deposition
charcoal beds	X			Org.	Coal		X				X		—	O	+	Incineration
galvanic sludge		X		Heavy met.	Heavy met.		X					X	+	—	O	Deposition
Process residues																
slag		X		Heavy met.	Slag		X			X			+	O	O	Deposition
distill. residues	X			Heavy met.	Org.		X		X				—	—	+	Incineration
Stored media																
batteries		X		Heavy met.	Heavy met.	X						X	O	O	—	Deposition
medicines			X	Org.								X	—	—	+	Incineration
Soil sanitation																
GKW-polluted soil		X		Org. Cl.	Soil			X	X				—	—	O	Deposition
dredged species		X		Org. Cl.	Clay			X	X				O	O		Discharge/deposition
				Heavy met.	Sand							—				

[1] —: No outlook for a technical/economic solution on a medium long-term basis. O: Outlook for a technical/economic solution on a medium basis. +: Existing technical/economic solution at hand.
Abbreviations: GWK's: chlorinated hydrocarbons. Org. Cl.: organic chlorine compounds. Heavy met.: heavy metals. Org.: organic compounds. C_xH_y: hydrocarbon compounds. Fe-verb.: iron compounds.

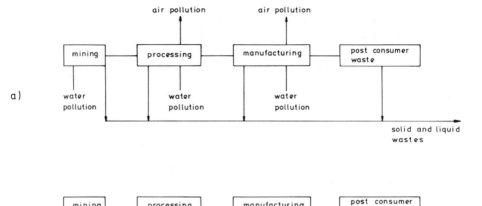
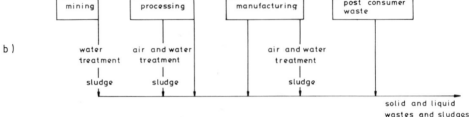

Fig. 1 — Production cycle in relation to environmental pollution and legislation. (a) and (b) both without waste legislation (a) without and (b) with clean air and water legislation.

treatment facilities could cope with in a proper and suitable manner, and the further production of xenobiotic substances. During the last decade there has been increasing public awareness that one of the major impacts made by technical development is the significant increase in the type, quantity and hazardous nature of the wastes being produced. It was also realized that there was a lack of technological know-how on disposal methods of 'tools' to effect management and control. This awareness was also stimulated by the recognition of the hazardous effects of some zenobiotic substances such as the polychlorobiphenyls, especially when it became clear that bioaccumulation occurs with disastrous effects on certain species. Promotion of this state of awareness is a process that takes a long time to develop and is increasing as the result of the shock-effect of the publicity given to specific incidents. Figure 2 attempts to visualize this process. Partly due to public pressure, legislation has been put forward to prevent problems arising from the generation and disposal of hazardous waste. It is not inconceivable that the present level of public awareness is nearing its zenith. Legislation and remedial activities concerning uncontrolled waste sites are also factors influencing this public awareness.

Control procedures to evaluate the effectiveness with which the legal requirements are carried out are essential, and these are beginning to show their worth in this direction. All of these measures can and will lead to a reduction of the type and amount of hazardous waste and an improvment in the methods of waste disposal by proper means of collection, transport, waste treatment and safe landfills. Cleaner

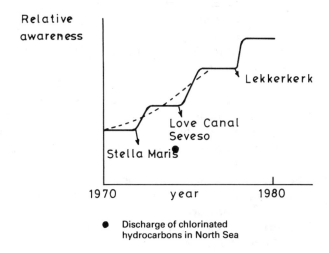

Fig. 2 — Visualization of the process of growing awareness.

technologies and 'good housekeeping' will materially assist in dealing with the problem as a whole. An attendant problem, however, is public opposition to the establishment of new industrial waste treatment facilities, particularly in or near populated areas. It has been estimated that it could take as long as ten years to establish a proper facility. A contribution to the solution of the problem might be to establish central waste treatment plants under a corporate body represented by both government and industry. An important effect accruing from this could be a decrease in public opposition to industrial waste-producing activities. Central treatment could also provide an essential contribution to the economic aspect of the problem of hazardous waste, particularly in respect of waste produced in small quantities at several places throughout the country.

5. THE MANAGEMENT OF HAZARDOUS WASTE

The management of hazardous waste is a fairly new discipline in most Western countries. In the late 1960s and early 1970s there was a general awakening of consciousness to the negative consequences for the environment and the danger to public health resulting from the careless, indiscriminate deposition of toxic and dangerous materials. In order to prevent damage to health and the environment by chemical wastes, it appears that one of the first conditions required is to provide adequate legislation to effect it and to stringently maintain its provisions. A second and equally important condition is that necessary disposal facilities are available.

Up until approximately ten years ago in Western countries, there was neither any experience of specific legislation in the field of hazardous wastes nor any appreciable experience with other methods of disposing of chemical wastes other than (uncontrolled) deposition of such wastes and their discharge at sea. The relatively rapid development of legislation and the progress already made with respect to the achievement of new disposal methods and facilities over the last few years are jointly due to international studies made by, or carried out on the orders of, the European

Economic Community (EEC), the Committee of the 'Challenges to Modern Society' of the North Atlantic Treaty Organization (NATO-CCMS), the Waste Management Policy Group of the Organization for Economic Co-operation and Development (OECD), the World Health Organization (WHO) and the United Nations Environmental Program (UNEP).

At national level however, there is a lack of sufficient insight into the quantities of waste involved and the disposal methods employed. Statistical data are often incomplete, unreliable or completely absent. An extra difficulty in making a comparison between different countries is introduced through the differing interpretations given by the various countries for 'waste substances' and 'chemical waste'. The whole matter is further obscured by whether or not secondary raw materials should be included, such as oils or solvents resulting from regeneration, or chemical wastes that are disposed of by the producer.

One of the most important developments in the field of hazardous wastes in Europe has been the establishment on 20 March 1978, of the EEC guidelines (78/319/EEC) relating to toxic and dangerous wastes. Together with a number of other guidelines (namely those relating to PCBs and PCTs, TiO_2 waste, used oil and those relating to wastes in general), these guidelines have proved a great stimulus in the development of legislation in the field of hazardous wastes in the member countries. They were obliged to bring their national legislation and the stipulations laid down therein into uniformity within two years of the guidelines coming into force. All member countries, with the exception of Greece, now satisfy the requirements laid down, while there remains considerable differences between the member countries in the matter of their interpretations and approaches to these requirements within the framework of the guidelines.

The guidelines have a legally-binding regulation on the 'cradle to grave' control of hazardous wastes by means of a licensing system and an obligatory reporting and documentation requirement amongst others.

A more extensive consideration of the comparison of a number of different aspects of legislation relating to the disposal structure is given in the Appendix.

6. POSSIBLE SOLUTION TO THE PROBLEMS CAUSED BY HAZARDOUS WASTE

Before presenting several methods of treating hazardous waste, it is once again necessary to indicate the range of priorities for these methods of disposal: prevention; recycling; processing and ultimate disposal.

6.1 Prevention
The possibility of preventing the existence of hazardous wastes is determined to an important degree by the raw material and energy aspects, the state of development of the techniques employed, economic and environmental considerations and the knowledge available.

Considering the point at which preventive intervention should be made in production processes, starting with raw materials, such as ore, the possibilities of prevention are limited. In addition to one or more valuable materials for which the production process is carried out, the ore will also contain other materials. Owing to their low concentrations, or for other reasons, these other materials are destined to become waste as a result of the ore-separating process. In such cases these factors prove to be an impediment to the amount of polluting substances which can be disposed of and to finding a suitable, effective method. An example of a polluting substance liberated in the production of zinc is jarosite.

While the choice of natural raw materials is limited, the choice of other materials is wider or can be made wider by research and development work. In a number of cases, for example, it has appeared possible to replace the application of cadmium in the metal-plating industry by the application of other materials less detrimental to the well-being of the environment.

Change(s) to an applied technique can lead in a number of cases to a reduction in the quantities of liberated substances. Examples of such changes are the 'load-on-top' system employed in the cleaning of tankers for soil transport and a method, still undergoing development, for discharging tankers in which, through the efficient use of discharging hoses and pumps, the amount of oil remaining aboard is further limited. In the latter case, residual amounts of oil remaining after discharge of the bulk often resulted in the pollution of considerable amounts of water used to wash out the tanks and through this change such pollution is further limited. Yet another example of a change in technique through which a considerable saving in energy can also be achieved is that in which chlorine is now produced from rock salt brine by electrolysis. Some of the older processing techniques were the cause of mercury emissions or an asbestos waste problem. By changing over to the use of membrane technology, it is now possible to produce chlorine without these problems arising. A start was made in 1984 with the application of this membrane technology in the Netherlands in a chlorine-producing plant which to date is the largest in the world. (Changeover to this technique is, however, time-consuming and involves a vast amount of capital expenditure.)

From the economic point of view it may appear that modifications to processes or techniques, perhaps carried out on a laboratory scale, are not viable on a practical scale or on economic grounds alone. In this context the comparative state of international commercial competition in this field can play a significant role. In taking note of the open Dutch market, great results can be achieved almost exclusively in the field of prevention when they support international consultation and policy making matters. The EEC thus constitutes a useful framework for the promotion of these ideas.

In so far as environmental-hygiene considerations are concerned, it must naturally not happen that earlier abated pollutant discharges are returned to the soil, air or water when measures are being taken to prevent pollutants from hazardous waste from finding their way to the same destinations.

6.2 Application and re-use of hazardous wastes

One may speak of the 'application of hazardous wastes' when these wastes (or components thereof) are used as raw materials, or as such for a particular purpose. Such a purpose need not necessarily have anything to do with the original purpose for which the materials were destined (before they reached the waste stage). When the wastes or their components are used for their original purpose, one may speak of re-use.

Alongside the traditional form of re-use involving the recovery of metals, other forms of re-use have come into existence and the great increase in energy costs has also contributed to this. Furthermore, and under the influence of the laws relating to hazardous wastes, industrial concerns themselves have become conscious of the need to dispose of hazardous wastes in the proper, but more costly, manner due to the requirements laid down in these laws. Application or re-use often carried out under the concern's own management can lead to considerable cost-savings.

Table 3 — Important applications of hazardous waste in the Netherlands

Waste	Re-use	Other useful applications
Cargo/load residues	Generation by distillation	Fuel
Chlorine containing residues from the chemical industry	Generation by per-chlorification	recovery of HCl
Dirty solvents	Generation by distillation	Fuel
Half-waste residues	—	Fuel
Sulphuric acid and sulphur waste	Worked-up to provide sulphuric acid	
Used catalysts	Regeneration	Recovery of heavy metals
Chalk waste	—	Mortar, agricultural chalk, water purification
Blast-furnace gas products	As ore after separation of zinc/lead fraction	
Aluminium containing solutions	—	Water purification
Used Fuller's Earth	—	Brick manufacture
Waste from thermal zinc-plating processes	—	Zinc-white manufacture
Baryte containing waste	—	Liquid-ballast for shipping
Photographic process wastes	—	Recovery of silver

Table 3 gives a summary of the most important applications of hazardous wastes in the Netherlands. Separating at source and recycling are extremely well-demonstrated in the electro-plating industry. Using proper design of the plating and rinsing processes combined with ion-exchangers, the formation of heavy metals-containing hydroxide sludges can be avoided and the pure metal salts can be reintroduced into the process system.

6.3 Processing
Processing methods are based on chemical, physical and/or thermal principles. The currently used techniques are:

(1) Chemical/physical detoxification
(2) Emulsion separation;
(3) Incineration in a rotating drum; and
(4) Incineration in a special incinerator (chlorinated hydrocarbons or pyrolizator).

Flow diagrams of some of these processes are given in Figs 3, 4, 5 and 6.

6.4 Ultimate disposal
Among the criteria used for determining the method of ultimate disposal, those relating to the environment are the most important. This implies that dispersion to the environment must be made impossible over a very long period of time (>30 years). Three systems of ultimate disposal are known: Deposition in specially adapted salt mines; deposition in specially designed containers, sealed at both top and bottom in a double-safety system; immobilization of the hazardous waste in binders.

7. HAZARDOUS WASTE TREATMENT COSTS
The methods of treatment of hazardous waste and their costs are summarized in Table 4.

Table 4 — Disposal methods and costs

Disposal method	Disposal costs ($/100 kg)
Final disposal (secured landfills)	10–100
Treatment (processing, incineration)	25–250
Re-use, recycling	50–500
Reduction, substitution (clean technologies)	?

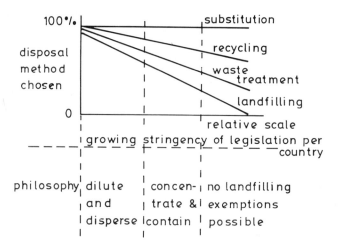

Fig. 3 — Relation between disposal method and stringency of legislation.

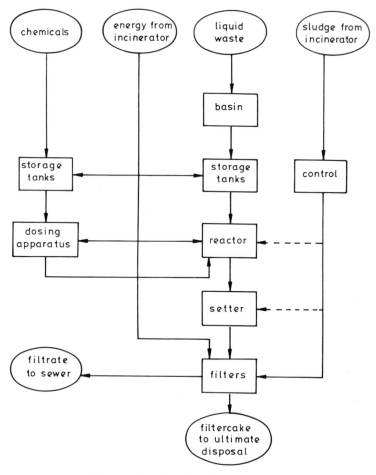

Fig. 4 — Flow sheet detoxification plant.

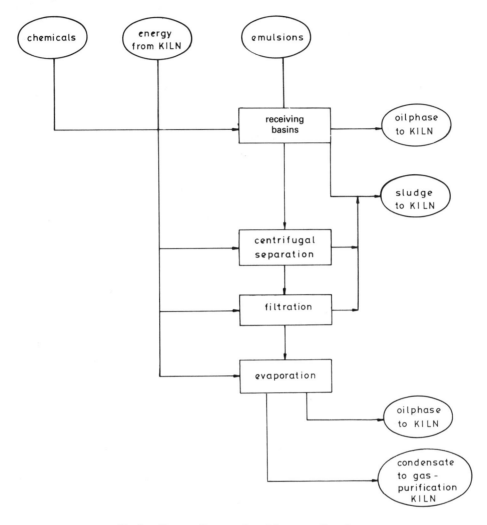

Fig. 5 — Process diagram of emulsion separation plant.

In general, the priorities of waste disposal options run contrary to the costs of the methods of disposal themselves. In other words, when a country's policy is to reduce final disposal costs as much as possible it has to assume in general that final disposal is the cheapest method. It is evident that when no legal restrictions exist on the manner of disposal, landfilling will be the most frequently applied method. If we take the legislation with the provision 'Prohibition of landfilling (with possibilities of exception)' as being the most stringent and 'No prohibition of landfilling' as being the least stringent, we might illustrate such conditions as in Fig. 7.

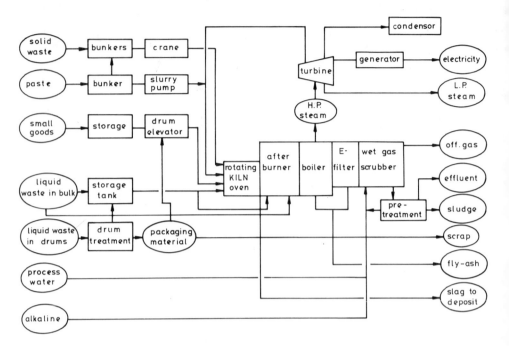

Fig. 6 — Process diagram of rotating kiln oven.

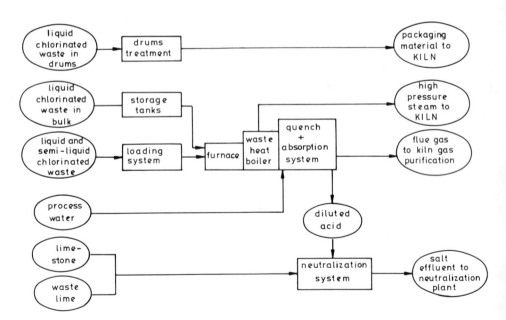

Fig. 7 — Process diagram of incineration of chlorinated wastes.

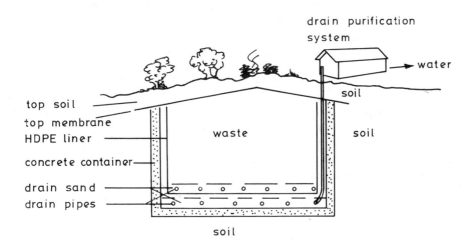

Fig. 8 — Utimate disposal of hazardous waste.

Table 5 — Summary of expected disposal costs (per ton of waste)[a] at the AV-Rotterdam facility

Type of waste	With maximum use of energy
Drum incinerator	Df 425
Chlorinated hydrocarbons	Df 634
Emulsions	Df 73
Wastewater distillation	Df 56
Detoxification/neutralization	Df 92

[a] 1 US dollar = Df 3.50.

8. ADVANCES IN RESEARCH AND DEVELOPMENT

One of the constraints placed on research and development in hazardous waste treatment methods and processes is the development of technology for the long term. The impact made by regulations and governmental policy are of importance in the exercise of research and development programmes.

Because of the differences in the approaches taken by the industrialized countries in tackling the hazardous waste problem it might be to their mutual interest to form a consensus of opinion on topics such as:

— the definition of hazardous wastes;
— the criteria and conditions for an environmentally safe, secured landfill, with due regard to the long-term risks involved;

— control procedures
— the improvement of standards of hazardous waste disposal technologies;
— substitution, prevention and recycling;
— research and development programmes.

The Dutch research and development programmes in this field are:
— The development of a system to remove zinc from blast-furnace produced substances and use the residue as a raw material in the production of steel.
— A laboratory-scale investigation into the extraction of chlorine from chlorinated hydrocarbons through which the hydrocarbons obtained can be used as a fuel.
— An investigation into the separation of wastes containing heavy metals into useful raw materials (metals, metal salts).

9. SOME CONCLUDING REMARKS

In summary, it can be said that:
— The definition and interpretation of the term 'Hazardous Wastes' differs from country to country so that a qualitative comparison between countries becomes difficult.
— At present in the case of Western industrialized countries, adequate legislation concerning hazardous waste has become either statutory or is in the process of development.
— The main route for disposing of hazardous waste in most countries consists of depositing it either on or into the soil (deposition).
— Observance of the governmental legislation on hazardous waste is insufficient in many countries through lack of manpower and the lack of control on the required observance of such legislation.
— Advanced disposal facilities are mainly provided with governmental financial assistance in investment and disposal costs alongside a strict observance of the legislation (good examples of this are evidenced by the situations existing in Denmark and the federal state of Bavaria, FRG).
— Re-use, useful applications etc., can be achieved on a large scale only if the government concerned creates the appropriate financial peripheral conditions.
— Prevention or minimalization of the existence of hazardous wastes is still not yet compulsory in any single country with the exception of several substances over which a consensus of opinion has been reached (PCBs and dioxins in this case).
— Transport of hazardous wastes from one country to another is an accepted phenomenon for many countries since often the opportunity for careful and economic disposal in their own country is not possible.

Table 6 — Selection and priority for R & D in the Netherlands

	Selection criteria				Area of attention				Priority						Comments
	A	B	C	D	E	F	G	H	I	J	K	L	M	N	
Organic wastes															
GKW's	O	+++	O	++	x				++	+	+	O	−	++	Two investigations in progress catalytic/thermal dechlorination
Waste waters	+	+	+	++		x			++	++	+	O	O	+++	One investigation in progress damp-air oxidation
Carbon beds	O	+	O	+	x				+	+	O	−	−	O	No investigation programmed
Distilled residues	O	+	O	+	x			+	+	−	−	−	−		No investigation programmed
Inorganic wastes															
Pickling baths	+	+++	O	+++	x				+++	+	−	−	−	+	No investigation programmed
Jarosite	+	+	+	++		x			++	++	O	O	+	+++	Belongs to the programmed investigation 'Detoxification of bulk wastes'
Shot-blasting grit	+	+	+	++	x				++	+++	+	O	O	+++	Investigation in progress
Calcium hydroxide sludge	+	+	+	++	x				++	++	−	O	+	+++	Belongs to the programmed investigation on 'Detoxification of bulk wastes'
Galvanic sludges	O	+++	+	+++		x			+++	++	O	O	−	++	First investigation phase closed on 'Separation by means of ion-exchange'
Slag	+	+	+	+++		x			++	++	+	O	+	+++	Belongs to the programmed investigation on 'Detoxification of bulk wastes'
GKW-polluted ground	+	++	+	+++		x			+++	++	+	O	+	+++	Follow-up investigation on thermal dechlorination
Dredged species	+	+	+	++		x			++	++	+	+	O	+++	First investigation phase closed on 'Cleaning of dredged species'
Batteries	O	+++	+	+++			x	x	+++	+++	+	+	O	++++	Investigation in progress on 'Collection, separation and processing of batteries'
Mixtures															
Shredder	+	+	+	++		x			++	++	+	O	O	+++	Investigation closed: result not (yet) applicable on economic grounds
Medicines	−	+	O	O	x				O	+	−	−	−	−	No investigation programmed

Source IMP Chem. A.

Explanation of column headings
Selection criteria: A=Quantity; B=Loading substances; C=Current disposal method; D=Desirable investigation
Area of attention: E=Prevention; F=Application/re-use; G=Process; H=Disposal.
Priority: I=Policy relevance; J=Environmental effect; K=Possibility of slag; L=Realization time; M=Cost of investigation; N=Priority.

Explanation of symbols

		Score	Criterion for score
Selection criteria	— Quantity	–	<1000 tons/year
		O	<50000 tons/year
		+	>50000 tons/year
	— Loading substance	+	<1% class A, B substances all corrections class C, D substances
		++	<5% class A, B substances
		+++	>5% class A, B substances
	— Present disposal	+	Deposition, discharge
		O	Destruction
		–	Application
	— Desirable investigation		Sum of the scored quantities, loading substance, present disposal.
Priority criteria	— Policy relevance		Sum of the scored quantities, loading substance, present disposal.
	— Environmental effect	O	No improvement in the disposal methods
		+	Prevention in place of application or processing
			Application in place of processing
		++	Processing in place of deposition/discharge
		+++	Application, prevention in place of deposition
	— Chance of success	–	Small
		O	Reasonable
		+	Good
	— Realization time	–	More than 10 years
		O	5 to 10 years
		+	0 to 5 years
	— Cost investigation	–	More than ƒ500.000
		–	Between ƒ200.000 and ƒ500.000
		O	Between ƒ100.000 and ƒ200.000
		+	Less than ƒ100.000
	— Priority		Half the sum of the policy-relevant environmental-effect, chance of success, time of realization and cost investigation.

BIBLIOGRAPHY

EEC Directive 78/319/EEC.
Hofman Commission Report VAR 1982.49 (in Dutch).
Kolfscholen Commission Report VAR 1982.21 (in Dutch).
Report on Chemical Wastes ISBN 90 047668.

APPENDIX

Comparison of a number of different aspects in the legislation and disposal structure
(Source: Appendix to IMP chemical wastes.)
In order to obtain a good survey of agreements and differences between various countries in matters of legislation and the disposal structure relating to hazardous wastes, a comparison has been drafted in an attempt to present the available information in tabular form. This abbreviated form of presentation has the advantage that a large amount of easily surveyed information can be displayed. The disadvantage is naturally that there are irrevocably many nuances in agreements in differences between legislation and disposal methods for hazardous wastes that get lost. We have endeavoured to remove this disadvantage by appending footnotes to the tables presented, where possible.

In the last ten years, legislation concerning hazardous wastes has become statutory in most industrialized countries. As has been said, the achievements made in establishing the necessary guidelines within the framework of the EEC has made a great contribution to the objectives within Europe. The guidelines so established have now been implemented by nearly all the member countries of EEC (with the exception of Greece). Legislation is now statutory in Sweden (1976), Finland (1978), Norway (1981), United States (1980) and Japan (1976). Legislation is still being drafted in Canada, amongst other countries, where in 1980 the 'Transportation of Dangerous Goods Act' became statutory and to which there will be added further implementation decrees. In Switzerland at the moment a draft decree is in preparation for supplementation to the statutory act of 1979 relating to environmental protection and in Austria, where up until now only the statutory act of 1979 relating to used oil has been in force. 'Cradle to grave control' exists to some degree in the Netherlands, United Kingdom and the Federal Republic of Germany and will also exist in Canada after implementation of the (draft) legislation.

A separate regulation concerning used oil is in force in most countries, whether it be included under the general regulations relating to wastes, in regulations relating to chemical wastes or as an entirely separate piece of legislation.

The main difference in legislation with respect to 'normal' (chemical) wastes is that a collection system is in force for used oil (in Denmark, France, the Federal Republic of Germany, the Netherlands and the United States amongst others).

A number of general aspects relating to legislation are given in Table A1. Legislation relating to the conditions and regulations posed in the incineration and discharge of hazardous wastes at sea has been enacted in many countries as the result of international agreements to protect the sea environment against pollution (the Treaties of Oslo, London, Paris and Helsinki).

In most countries, hazardous wastes are defined with the aid of a list of the wastes

Table A1 — Legislation relating to hazardous wastes

	Belgium[a]		FRG	Denmark	France	Netherlands	UK	USA
	Nat.	Vl.						
Legislation	+	+	+	+	+	+	+	+
Part of the general legislation relating to municipal wastes	−	+	+	−	+	−	+	+
Year of legislation	1974	1981	1972	1972	1976	1976	1974	1976
Year indicating hazardous wastes	1976	−	1977	1976	1977	1979	1980	1980
Implementation of legislation	+	−	+	+	+	+	+	+
Legislation relating to used oil	−	+	+	+	+	+	−	+
Separate enactment decree within main legislation (hazardous wastes) relating to used oil	−	+	−	+	+	+	−	+
Mainly with enactment of legislation relating to the official loading level[c]	N	D	D	G	D	N	G	D
Special legislation for disposal and incineration of (hazardous) wastes at sea	−	+	+	+	+	+	+	+
'Cradle to grave' control for hazardous wastes	−	−	+	+	−	+	+	+[d]

[a] In Belgium, a distinction must be made between National legislation and legislation concerning the Flanders region alone. This distinction is made in the table by the annotations National/Flemish. Specific legislation relating to chemical wastes in the Wallonian region has not yet been enacted.
[b] The 'Deposit of Poisonous Waste Act' was withdrawn in 1972 and replaced by the 'Special Waste Regulations 1980'.
[c] N (National Federal); D (Federal State, Department, Province); G (Municipal, Regional).
[d] Reporting pro-formas (trip-ticket) are not submitted to EPA unless the producer, transporter or processer signals a deficiency in pro-forma flow. However, in many states there is an obligatory duty to report.

to which the legislation is applicable. Lists of wastes are used in the Federal Republic of Germany, France, the Netherlands, Belgium, Denmark, Sweden, Finland, the United Kingdom and the United States amongst others. The EEC guidelines relating to toxic and dangerous wastes also has a list of such wastes. There is a certain degree of agreement between lists; however, very great differences can be ascertained. For example, there are wastes which have a re-use value in some countries which do not fall within the legislation (Federal Republic of Germay, Norway) and which in some countries are also quoted as being industrial processes (the Netherlands, Federal Republic of Germany) alongside the wastes in the lists and which are jointly determinative as criteria for toxicity (United Kingdom, United States) and concentrations (Belgium, the Netherlands). Table A2 gives a comparison between different countries.

A number of differing starting-points for legislation and their achievement in practice are further considered in Table A3, and particularly an assessment of the 'polluter-pays principle' on the basis of government financial participation in the provision of the costs of definite disposal and the intentions of the legislation with regard to re-use, useful application and their promotion in practice. In nearly every country, the government contributes to the disposal costs to a greater or lesser degree (whether it be through an investment contribution, by subsidizing the disposal costs, by means of interest-free loans or by underwriting losses).

In the United Kingdom and the United States, it is the 'polluter' who bears most of the costs of disposal; however this is certainly not the case in Denmark. In France, and to a lesser degree in Belgium and the Federal Republic of Germany, there is a substantial degree of State contribution towards disposal costs.

Re-use and useful application of hazardous waste can often only be achieved if recourse to this results in certain economic advantages. In many countries re-use is stimulated by the state by initiating research, by propaganda and sometimes through holding a waste exchange, and by technical and financial assistance as examples.

Advanced disposal methods such as incineration and physical/chemical treatment (e.g. detoxification) and the different forms (costly) of making waste suitable for re-use (regeneration) and useful applications only reach a good state of development in countries where there is a possibility of limiting deposition of wastes onto or into the soil (either by legislation or through the hydrogeological situation).

The United Kingdom is a good example of a country in which deposition (whether or not controlled) is the main disposal option. Other disposal methods cannot compete financially with this method. In countries where there is a large degree of state intervention in the disposal structure (Denmark and some of the states of the Federal Republic of Germany, such as Hesse and Bavaria), advanced developments have also been made in respect of disposal facilities as the result of stringent legislation which opposes certain disposal options (levies, reporting obligations) or exclusion (export ban), and in any case by means of large financial contributions by the state.

Several aspects of legislation relating to hazardous waste disposal have been given in Table A4 and in Table A5 an attempt is made at a quantitative estimate of the different disposal methods employed in the countries under consideration. The figures given in the tables have been derived from results of inventories made by OECD, from inventories within the framework of the Treaties of London and Oslo

Table A2 — Definition of hazardous wastes in legislation and enactment decrees

	Belgium	FRG	Denmark	France	Netherlands	UK	USA
(Waste) list	+	+	+	+	+	+	+
List of processes	−	+	−	+	+	−	−
Concentrations	+	−	−	−	+	−	−
Criteria (e.g. toxicity, flammability, corrosiveness)	+	−	−	−	−	+	+

Table A3 — Aspects of legislation on hazardous wastes and the achievement of its objectives in practice

	Belgium	FRG	Denmark	France	Netherlands	UK	USA
The 'polluter pays principle'[1]	+	+	+	+	+	+	+
State contribution to the costs of disposal[1]	O	O	+	+	O	–	–
Intentions regarding prevention/minimization	+	+	+	+	+	+	+
Possibilities to force prevention/minimization by legislation	–[2]	–	–	+/–	+/–	–	–
Intention to re-use/usefully apply	+[2]	+	+	+	+	+	+
Control possibilities for re-use/useful application in the legislation[3]	–	+/–[4]	–	–	+/–	–	–
Promoting re-use, useful applications, prevention and minimalization through:							
State establishment of a special authority charged with handling these matters	–	–	–	+[5]	–	–	–
State action to carry out and stimulate research by making financial provisions available	+	+	+	+	+	+	+
Financial support by the State in actual activities	–	+/–	+/–	+/–	–	–	–
Establishment of a waste exchange	–	+	–	+	+	–	+/–[4]

[1] –, No contribution, O, relatively small contribution; +, relatively large contribution.
[2] Only in the Flemish region up to now.
[3] These means are enacted in principle in various National legislations, but rarely or never used in practice.
[4] Diverse regulations in the different countries/Federal States.
[5] A special institute exists in France for handling these matters (ANRED).

Table A4 — Several aspects of chemical waste disposal

	Belgium	FRG	Denmark	France	Netherlands	UK	USA
Public nuisance obligations (in which the government decides whether the responsibility for the wastes is taken over by the state when the producer reports their presence)	−	+	+	−	−	−	−
Disposal facilities in the hands of the state	+/−[a]	+/−[c]	+	−	+/−[b]	−	−
Important state financial contribution toward the costs of disposal	−	−	+	+	−	−	−
Levies of taxes imposed on (disposal of) hazardous wastes	+	+	−	+	+	−	+/−[d]
The obligation of the state to set up a plan for the disposal of hazardous wastes[f]	+[e]	+	+	+	−	+	+

a, b, c, d, e — see footnote to Table A3.

Table A5 — A very rough quantitative estimation of the methods used for the disposal of home produced chemical wastes (in % excl. used oil, 1980–1983)[1]

	Belgium	FRG	Denmark	France	Netherlands	UK	USA
Deposition	4	40	7	27	16	68[3]	85
Incineration	n.b.	10	57	5	21	2	2
Physical-chemical treatment (inc. immobilization)	n.b.	10	14	3	13	13	2
Re-use	n.b.	3	n.b.	1	4	7	3
Incineration at sea	1	2	0	<1	2	<1	n.b.[2]
Discharge at sea	45	36	14	61	1	10	6
Export	2	<1	1	<1	25	<1	<1
Total (in millions of tons)	1.1	6.2	0.1	5.4	0.5[7]	4.4	42.0

[1] The figures quoted in the related literature are contradictory. In order to establish the quoted numbers, telephone contact was made with representative of the various authorities concerned.
[2] As far as is known, new activities are being started up in the USA.
[3] Disposal under their own management is of great importance (certainly 50%).

and from various related publications in international literature. In spite of all this, there still exist some uncertainties relating to the figures given in the table, mainly due to the different national interpretations of the meanings of the terms 'Hazardous wastes', 'reuse' and 'recycling' for example. Quite often only 'off-site' disposal is referred to in the literature. The degree to which hazardous wastes are disposed of by the producer can only be an approximation.

Part 6
Environmental policy and regulations

6.1

The EEC directive on environmental assessment and its effect on UK environmental policy

Brian D. Clark
Executive Director, Centre for Environmental Management and Planning, University of Aberdeen

1. INTRODUCTION

There are many who would argue that the EEC Directive on Environmental Assessment is an event which will have a major impact on environmental policy within Britain in the coming years. This chapter, based on work undertaken by the Centre for Environmental Management and Planning for a consortium of central and local government and industry, puts the Directive in context and considers how it relates to current and future British environmental evaluation procedures. It is suggested that much of the opposition to the Directive by certain government departments and some industralists is unfounded and that it will have considerable economic and environmental benefits.

2. BACKGROUND TO THE DIRECTIVE

The Directive is entitled 'A Directive Concerning the Assessment of the Environmental Effects of Certain Public and Private Projects'. The version under review is the latest of many drafts prepared in furtherance of the EEC Action Programmes on the Environment for 1973 and 1977. The Treaty Power on which the proposal relies is the Treaty establishing the EEC, Article 100. The Directive was approved by the Council of Ministers in Rome in March 1985 and Member States must take the necessary measures to comply with the Directive within three years.

3. PURPOSE AND EFFECT OF THE DIRECTIVE

The aim of the Directive is to introduce general principles for the prior assessment of environmental effects, with a view to improving the operation of planning procedures governing private and public activities which might have significant effects on the environment. Particular attention is drawn to planning and decision-making with respect to individual projects, land use plans, regional development programmes and economic programmes including those in particular sectors.

To avoid undue initial administrative burdens in Member States, however, principles of assessment are to be introduced step by step with priority being given to planning and decision-making procedures for authorizing projects. Due to the availability of procedures for the control of projects in all Member States, the Directive anticipates that the introduction of assessment principles will not require the development of new procedures.

4. EXAMINATION OF THE PROVISIONS OF THE DIRECTIVE

The provisions and procedural stages to be followed in the application of the Directive are summarized in Table 1 and details of the various articles are included in the Appendix.

5. INTEGRATION OF THE DIRECTIVE WITH EXISTING UK PROJECT AUTHORIZATION PROCEDURES

5.1 Existing UK project authorization procedures

Under existing procedures for project authorization in the UK, development projects are subject to a wide range of controls under which permission is needed from public authorities before they can be undertaken. The permissions required differ with the scale of the project, the activity or use for which it is intended, detailed design, location and other factors. Not all of these authorization procedures are concerned with the environmental effects of the proposed development. The two kinds of control of prime significance with reference to the Directive are planning and pollution control. Some others are of lesser or occasional significance.

Other procedures in which environmental considerations play some part are usually specific to particular classes of development. For example, all new power stations, transmission lines and oil and gas pipelines require an authorization from the Department of Energy, which carries with it deemed planning permission. Similarly, oil refineries require government authorization (licence) which must be obtained before a planning application can be accepted. This also applies to onshore oil drilling and production. Motorways and express roads are authorized by the Department of Transport, and new towns and water supply schemes by the Department of the Environment and the Scottish Office.

Public developments are subject to different authorization procedures than private developments. Developments by government departments and planning authorities are covered by circulars and are more variable depending on the nature of the proposal. For example, classes of the General Development Order given permitted development status to sundry minor operations carried out by British Rail, docks, canal and harbour undertakers, water supply undertakers, gas boards, electricity boards, tramway undertakers and the Post Office. Some operations are subject to the authorization of specific government departments under a variety of

Table 1 — Provisions and procedures in the application of the European community directive on environmental assesssment (EA)

Stage 1	Stage 2	Stage 3	Stage 4	Stage 5
Article 4	*Article 6*	*Article 8*	*Article 9*	*Article 11*
Projects listed in *Annex I* subject to *mandatory* EA	*Consultations* with the relative administrative authorities and other statutory bodies, and opinions obtained	*Information gathered under Articles 5–7* to be *taken into account* in reaching a decision by competent authority	When *decision taken* competent authority will make publicly known: — content of decision and any conditions — reasons and considerations on which decision based	Information and experience exchanged between member state in EEC, including criteria and thresholds adopted
Projects listed in *Annex II* subject to EA at the discretion of member states but based on application of criteria and thresholds	Information obtained is made available to public *Public consultations* to obtain opinions		Similar information sent to member states affected under Article 7	Commission reports on experiences of member states to the council and parliament after five years
Simplified assessments or exemptions from assessment subject to compliance with conditions	*Article 7* Where the project is likely to have *trans-frontier effects*, available information to be sent to member states likely to be affected as a basis for consultations		*Article 10* Limitations placed on competent authority by: — national regulations — administrative provisions — accepted legal practices respecting industrial and commercial secrecy and safeguarding the public interest These are respected	Subsequent amendments may be made to the Directive
Article 5 Where EA is required *developer* must supply *information* as specified in *Annex III* (see Table 4). Minimum required: — description of project site, design and size — data to identify and assess main effects on environment — measures to avoid adverse environmental effects — non-technical summary				*Article 12* Directive to be adopted three years after notification Texts of national laws to be sent by member states to Commission *Article 13* Member states may apply stricter environmental rules, if desired

acts and other regulations. Such authorization can and usually does carry with it 'deemed' planning approval. It is the responsibility of the developing undertaker to consult the planning authority and such representations as may be made by the planning authority are taken into account by the authorizing government department. On the other hand, there are operations by statutory undertakers that do require the approval of the planning authority in the normal way, for example, gas terminals and pipelines under 10 miles in length.

The essential characteristics of the UK development control system for the authorization of development projects are as follows:

(i) Existing procedures are comprehensive and no major development with significant environmental effects escapes some form of control.
(ii) Private and public developments are subject to broadly similar considerations although following different procedures.
(iii) The majority of decisions on major development proposals are taken by district or regional authorities in Scotland, but the most contentious may by one means or another come to be decided at central government level, under both planning and pollution controls. All oil- and gas-related development must be notified to the Secretary of State for Scotland.
(iv) In reaching decisions there is much local discretion in the criteria taken into account and the precise values attached thereto. There is minimal application of national standards and in the pollution field the concept of 'best practicable means' is firmly established.
(v) At some stage, particularly under planning control, the public has rights of information, participation and objection in relation to development proposals.

In summary, the boundaries within which planning authorities in the UK can operate in dealing with private applications for planning permission and proposals for development by public authorities are wide enough to enable them to take account of the total effect which a particular development is likely to have on its surroundings both immediately and in the longer term.

5.2 UK procedures and the Directive

The assessment and consultation procedures set out in the Directive can be broadly aligned with those of the UK development control system as shown in Table 2. Implementation of the Directive within three years will entail some changes in both subordinate and main legislation but it is unlikely that this will raise any difficulties at central or local government levels. Under the present Planning Acts and General Development Order, a planning authority can require a private developer to supply whatever information may be deemed necessary to enable the authority to determine a planning application. This would appear to be an adequate power for the purposes of the Directive. However, additional powers will be required to deal with public projects for government departments, statutory undertakers and planning authorities under the provisions of the Directive.

The planning process will not necessarily take longer than present, and the preparation of environmental assessments may well have a beneficial effect in expediting decisions and in the progress of public inquiries. The Directive, however,

Table 2 — Comparison between UK development control and EEC directive procedures

UK development control	EEC Directive
Preparatory activities	
Developer decides to proceed with project	Developer decides to proceed with project
Informal discussion with planning authority about relationship of project to the development plan for the area and what information required including an EA and what topics or key issues are to be covered	Initial screening process to ascertain status of project in relation to types of project listed in Annex I and Annex II of the Directive: — Is a mandatory assessment (Annex I), or a discretionary assessment (Annex II, based on the application of criteria and thresholds) required? — What type of assessment is required: full assessment, simplified assessment, or exempt?
Developer prepares planning application and may or may not decide to hold public meeting about project	Developer prepares planning application in accordance with the requirements of Articles 4 and 5, covering information listed in Annex III, with the assistance of the appropriate planning authority and agencies as may be necessary
Submission	
Developer submits formal outline or full planning application with supporting information (not a statutory requirement)	Developer submits formal outline or full application for planning permission together with such supporting and other information to satisfy the provisions of the Directive
Planning authority enters application in planning register and notifies Secretary of State and regional/county authority in case of oil- and gas-related developments under possible 'call-in' procedures. Notices put in press	Planning authority follows normal statutory procedures on receipt of planning application
Consultation	
Statutory and public consultations within time limits laid down	Consultations with relevant administrative authorities, and other statutory bodies with environmental interests and opinions obtained
(This information not necessarily made available or public opinion sought of proposal. Where there is strong public feeling there may be individual or pressure groups formed to voice opinions)	Information made available to public and opinions obtained through consultation
	Information sent to Member States that may be affected by transfrontier pollution.
Planning authorities carry out appraisal of considerations and prepare reports for appropriate committees. Depending on circumstances Secretary of State or regional/county authority may hold a public inquiry. District may hold a non-statutory hearing	Assessment made of the likely significant impacts based on information provided for under Articles 5–7.
Local Planning Authority Recommendations	
Recommendations made by planning committee and final decision taken by full council	Decision and conditions made public together with reasons and considerations on which the decision was based.
Decision may be refused and lead to an appeal or approved with or without conditions, or in the case of outline planning permission, approved with reserved matters to be dealt with later	Information and experience obtained in dealing with project EAs and criteria and thresholds applied sent to EEC

does require the developer to carry out work which might not otherwise be done, especially if the planning authority makes use of the Directive to require, at no cost to itself, quantified studies of aspects of the environment which it might not have looked into in such detail itself.

6. CONCLUSIONS

Comparison between the proposed Directive procedures and those of the UK development control system indicates a number of technical and administrative aspects of the Directive that need discussion, clarification and agreement, preferably through consultation between central government, local government associations (such as the COSLA and SSDP in Scotland) and industry associations (such as the CBI and UKOOA). The Department of the Environment has set up a working group to consider how the Directive should be implemented within the UK.

APPENDIX

Article 1. Definitions

This article states that the draft Directive applies to the assessment of the environmental effects of those public and private projects which are likely to have significant effects on the environment. The meaning of the terms 'project', 'developer' and 'development consent' are given and reference is made to the fact that projects serving national defence purposes are not covered by the draft Directive.

Article 2. Application of the directive and exemptions

The general principle is stated that without an appropriate assessment, consent cannot be given to projects likely to have significant effects on the environment by virtue of their nature, size or location. These projects are defined in Article 4.

Provision is made for Member States to integrate environmental assessment into the existing procedures for project consent or to establish new procedures to comply with the aims of the Directive.

In exceptional cases, Member States may, subject to compliance with certain provisions, exempt a specific project in whole or in part from the provisions laid down in the Directive (see Article 10). Provision is made for considering another form of assessment; making information available to the public; and prior to granting consent informing the Commission of the reasons justifying the exemption granted and the information available. The Commission will forward this information to other Member States.

Article 3. Scope of the assessment

The concept of 'environmental effects' is defined. The environmental assessments must identify, describe and assess the direct and indirect effects of a project on human beings, fauna and flora; soil, water, air, climatic factors and the landscape; the inter-relationship between them; and material assets and the cultural heritage.

To avoid the creation of unfavourable competitive conditions between Member

States, common principles are set out to harmonise the development control/decision-making procedures of the Member States, regarding:

— the main obligations of developers
— the types of projects to be subject to assessment
— the content of assessment
— the environmental features to be taken into consideration.

The Directive proposes:

(i) a general requirement to ensure that planning permission for projects should only be granted after an appropriate prior assessment of the likely significant environmental effects has been carried out;
(ii) co-operation between planning authorities and developers in providing appropriate information on the possible range of the environmental effects of a given project, as well as on the reasonable alternatives to it;
(iii) a general requirement to consult all statutory bodies with a responsibility for environmental matters, and other Member States about the trans-frontier effects of a project; and
(iv) a requirement to inform the public of the issues relating to particular projects and to provide an opportunity for their views to be made known — subject to normal constraints of national security and commercial confidentially.

Many revisions and amendments have been made to the fourteen articles and three annexes in the Directive to overcome objections in the earlier drafts and it now provides a more pragmatic and flexible approach to environmental assessment. The main provisions of the Directive are:

(i) mandatory assessment, in a broadly prescribed manner, of major development projects prescribed in Annex I;
(ii) assessment at Member States' discretion of projects listed in Annex II, on the basis of criteria and/or thresholds to be established to determine their environmental consequences; and
(iii) information listed in Annex III to be submitted by the developer to the responsible authority along with applications for project development approval.

Provision is made for Member States to determine whether projects of the classes listed in Annex II should be made subject to a simplified environmental assessment or to be exempted from any environmental assessment. No guidance is given as to the content or coverage of a simplified assessment.

Article 4. Form of assessment for projects of classes listed in Annex I and Annex II

The field of application of the Directive is clearly indicated. In Article 4(1) the classes of projects listed in Annex I are made subject to a mandatory 'full' assessment in accordance with Articles 5–10. The classes of projects listed in Annex II include projects which are not always likely to produce significant effects on the environment. In Article 4(2) provision is made for Member States to determine whether these projects should be made subject to a 'simplified' assessment or exempted altogether. The competent authorities will have to identify the projects which deserve 'full' assessment and those which should undergo 'simplified' assessment. This is to be achieved by specifying types of projects or by establishing criteria and/or thresholds for selecting projects.

Article 5. Data and information provision

This article lays down the first stage of the assessment procedure which involves the supply by the developer of the information needed to assess the environmental effects of the proposed project. The minimum required information is set out. In preparing this information, provision is made for the developer to receive available relevant information from the competent authority, while remaining repsonsible for providing the facts and figures required of him.

Reference is made to Annex III, which gives a breakdown of subjects which the information provided by the developer should cover. The developer's obligation to provide such information is not unlimited, but information supplied must be relevant to the given stage of the consent procedure, the project characteristics, and the likely environmental effects.

Article 6. Consultation and information

This article is concerned with the second stage of the assessment process which involves consultation between the various parties concerned. It requires identification of and arrangements for consultations with public authorities with specific responsibility for environmental matters. This consulation procedure is important since it may serve to coordinate the comments of various parties in respect of a specific development project.

The article also deals with making the gathered information available to the public and with obtaining their opinions before the project is approved. The detailed arangements for effective information dissemination and consultation are to be decided by the Member States, which may determine the public concern and specify the places and the way the information can be consulted, for example, by billposting, press advertisements and exhibitions. In addition, the Member States may determine the manner of consultation, for example by written submissions and public inquiry, and the time limits for the various stages of the procedure.

Article 7. Trans-frontier effects

This article provides for consultation on trans-frontier issues. Where a project is likely to have significant environmental effects in another Member State, the competent authority should send to the appropriate authorities of the Member State likely to be affected the information on the project provided by the developer under Article 5. This should occur at the same time as national consultation procedures. This information may be used as a basis for any future consultations. No right of veto on administrative decisions is given to Member States that may be affected.

Article 8. Incorporation of assessment into decision-making
The information received from the developer and the other information and comments received during consultation procedures must be taken into account in the development consent procedure.

Article 9. Publication of decision and reasons for decision
The competent authority is required to make public the contents of the decision and any conditions together with the reasons and considerations on which the decision is based. The arrangements for making this information available to the public are left to the Member State. Those Member States informed under Article 7 must also be notified of the decision.

Article 10. National regulations and legal practices
This article recognises that the provisions of the Directive shall not affect the obligation on competent authorities to respect the limitations imposed by national regulations and administrative provisions and accepted legal practices with regard to industrial and commercial secrecy and the safeguarding of the public interest. The limitations are also applicable to the provisions for transfrontier issues under Article 7.

Article 11. Exchange of experiences and commission review
This deals with the exhange of information between Member States and the Commission on experience gained in applying the Directive and on the criteria and/or thresholds adopted for the selection of projects for assessment. Five years after the notification of the Directive, the Commission will submit to the Council and European Parliament a report on the application and effectiveness of the Directive based on the exchange of information between Member States. At that time, the scope and procedures of the Directive will be reviewed and recommendations made to improve the assessment methods.

Article 12. Implementation time
Measures to comply with the Directive must be taken within three years of its notification, and the Commission provided with texts of the provisions of national law to meet the requirements of the Directive.

Article 13. Stricter national procedures
Provision is made for Member States to apply stricter rules regarding scope and procedures when assessing environmental effects.

Article 14. Greenland
The Directive does not apply to Greenland.

ANNEX I. PROJECTS SUBJECT TO 'FULL' ASSESSMENT UNDER ARTICLE 4(1)
This lists the classes of projects subject to 'full' assessment, as set out in Articles 5–10, whatever their size or site location. Assessment is regarded as mandatory for any new

project falling within the classes set out in this Annex. A list of oil- and gas-related projects in Annex I is given in Table A1.

Table A1 — Extract from EEC Directive

Annex I

Oil- and gas-related projects subject to Article 4(1)

1 Crude-oil refineries (excluding undertakings manufacturing only lubricants from crude oil) and installations for the gasification and liquefaction of 500 tonnes or more of coal or bituminous shale per day
2 Integrated chemical installations
3 Trading ports and inland waterways and ports for inland-waterway traffic which permit the passage of vessels of over 1350 tonnes

ANNEX II. PROJECTS SUBJECT TO ASSESSMENT UNDER ARTICLE 4(2)

This lists the classes of projects regarded as capable of having significant environmental effects under certain conditions deriving from their own characteristrics, for example, when they are larger than a given size or cause a given amount of pollution. The competent authority is required to consider which of these development projects should be subject to a 'full' assessment within the meaning of Articles 5–10. To this end, there is a need to determine the criteria for fixing the technical thresholds (size, production, emissions, etc.) or financial thresholds (construction costs, etc.) beyond which a project becomes subject to an assessment and finally, on the basis of those criteria and thresholds, actually select the projects. Within this approach, the competent authority may determine whether or not a full or simplified assessment is required or none at all. A list of oil- and gas-related projects in Annex II is given in Table A2.

ANNEX III. INFORMATION TO BE PROVIDED BY THE DEVELOPER UNDER ARTICLE 5(1)

This Annex sets out the information the developer must submit to the competent authority along with his planning application. The most important features are those describing the environment likely to be affected; those assessing the most significant effects of the proposed project; and those explaining the reasons for choosing the particular project and/or site among other alternatives which might reasonably have been considered. Information on the alternatives considered should enable the various authorities and public consulted, to make a choice representing the best compromise between environmental considerations and other interests at stake. In this regard, it would be unreasonable to except the developer to provide a complete decription of the various alternatives available, if any. The information referred to in Article 5 and Annex III is set out in Table A3.

Appendix A2 — Extract from EEC Directive

Annex II

Oil- and gas-related projects subject to Article 4(2)

1 *Extractive industry*
   — deep drillings
   — extraction of petroleum
   — extraction of natural gas
   — extraction of bituminous shale
   — surface industrial installations for the extraction of petroleum, natural gas and bituminous shale

2 *Energy industry*
   — industrial installations for carrying gas
   — surface storage of natural gas
   — underground storage of combustible gases
   — surface storage of fossil fuels

3 *Chemical industry*
   — Treatment of intermediate products and production of fine chemicals (unless included in Annex I)
   — Production of pesticides and pharmaceutical products, paint and varnishes, elastomers and peroxides
   — Storage facilities for petroleum, petrochemical and chemical products

4 *Infrastructure projects*
   — Construction of roads, harbours and airfields not listed in Annex I
   — Pipeline installations

5 *Other projects*
   — Installations for the disposal of industrial and domestic waste included in Annex I
   — Water-purification plants
   — Sludge-deposition sites

6 *Modifications and testing*
   — modifications to development projects included in Annex I
   — projects in Annex I undertaken exclusively or mainly for development and testing of new methods or products and not used for more than one year

Table A3 — Extract from EEC Directive

Annex III

Information referred to in Article 5(1)

1. Description of project, including:
   — the physical characteristics of the whole project and the land-use requirements during the construction and operational phases;
   — the main characteristics of the production processes, i.e. nature and quantity of the materials used;
   — an estimate, by type and quantity, of the expected residues and emissions (water, air and soil pollution, noise, vibration, light, heat, radiation, etc.) resulting from the operation of the proposed project.

2. Where appropriate, an outline of the main alternatives, studies by the developer and an indication of the main reasons for his choice, taking into account the environmental effects.

3. A description of the aspects of the environment likely to be significantly affected by the proposed project, including, in particular, population, fauna, flora, soil, water, air, climatic factors, material assets, including the archaeological heritage, landscape and the inter-relationship between the above factors.

4. A description including the likely significant direct effects and any indirect, secondary, cumulative, short, medium and long-term, permanent and temporary, positive and negative effects of the project on the environment resulting from:
   — the existence of the project
   — the use of natural resources
   — the emission of pollutants, the creation of nuisances and the elimination of waste, and

   the description by the developer of the forecasting methods used to assess the effects on the environment.

5. A description of the measures envisaged to prevent, reduce and where possible offset any significant adverse effects on the environment.

6. A non-technical summary of above information.

7. An indication of any difficulties (technical deficiencies or lack of know-how) encountered by the developer in compiling the required information.

6.2

UK pollution control — legal perspectives

Richard Macrory,
Imperial College, London

In an ideal world the opportunities presented by developing technologies in the field of environmental management would be rapidly and gratefully embraced by relevant legal controls. In practice this may rarely be the case. The need to accommodate a range of principles and interests in the design of legislation and its interpretation over and above the effective implementation of technical solutions may produce results that sometimes represent uneasy compromises, and at the very least cause the law to lag behind available technology. In the foreseeable future, legal controls will continue to play an essential role in the implementation of policy initiatives, and this chapter therefore approaches the issues of pollution abatement and management from a legal and administrative perspective. The detailed observations relate to the British experience, and there is a danger in translating the issues raised to other jurisdictions since legal and administrative practice and tradition often bear strong, but sometimes subtle, connections to a purely national cultural and social framework. Although countries may share pollution problems and the availability of technological solutions, it does not follow that the legal and administrative issues raised will be similar.

Recent experience in Britain suggests the emergence of four general themes which I believe will play a critical role in the design of future pollution control policy, and which will place demands on those charged with legislative design. I have classified these themes as (i) *anticipation,* which implies a non-reactive mechanism of control allowing prior identification of future pollution loads and types; (ii) *integration,* which implies the need to consider the inter-connections between possible control strategies, particularly where these cross a medium or particular pathway of pollutant, in order to ensure that the trade-offs involved in adopting one strategy in preference to another are properly understood and taken into account; (iii) *participation,* which concerns the involvement of the general public, or perhaps more accurately third parties, in a process of decision-making which might otherwise rest largely on a two-way relationship between a regulatory agency and the 'polluter';

and, finally, (iv) *explicitness*, by which is meant the extent to which pollution standards or goals receive overt expression in legislative instruments.

In the following sections of the chapter, the adequacy of current British pollution law in respect to these four criteria is considered, and an inconsistent picture is revealed. Emerging legislation and policy, however, can be expected increasingly to reflect them, though their accommodation will not necessarily assist the implementation of appropriate technical solutions to pollution problems.

2. ANTICIPATION

2.1 The planning system

The 1947 Town and Planning Act introduced a comprehensive land-use planning system on a national basis to Britain with a result that from that date, 'ownership of land, generally speaking, carries with it nothing more than the bare right to go on using it for existing purposes' [1]. Given that the legislative provisions provide extensive powers to public authorities over private rights, they are remarkably unspecific as to the precise purpose of the powers. Perhaps deliberately so, for the flexibility in the drafting of the legislation has allowed the planning system to develop from a concern with land-use and amenity in a narrow sense to encompass a wide range of social and economic considerations — admittedly a broadening of scope that has largely been initiated by the planning profession bringing in their wake a sometimes less than enthusiastic judiciary [2]. Planning powers under the Town and Country Planning legislation are *par excellence* tools of anticipatory management, and in the context of pollution control have an obvious importance.

The term, 'pollution', though, is not to be found in the primary legislation. The closest explicit reference is to be found in the provisions defining subject matter to be included in Structure Plans, the required broad-based land-use plans providing a framework for the taking of individual development control decisions. The Structure Plan Statement must include, 'measures for the improvement of the physical environment' [3], terminology which, according to Central Government advice, includes land-use policies designed to minimize pollution. Such plans have a clear role in such matters as locational priorities for industrial use with pollution implications, but may create problems where policies trespass too directly on the interests of specialized pollution control agencies. The most notorious and clear example of such a conflict occurred in 1977 when Cheshire County Council proposed for inclusion in their Structure Plan planning policies concerning new developments likely to cause a deterioration of ambient air quality standards (which had no legal status outside their inclusion in the Plan) which were expressly based on WHO targets [4]. Central Government rejected the proposed policy on the grounds that it would be inappropriate, 'to approve policies based on standards which have not yet been accepted nationally and which are likely to give rise to difficulties in implementation'. The reference to quantitative standards was therefore excised, but three years later, as a result of an agreed EEC Directive, air quality standards concerning SO_2 and suspended particulates became part of British law [5]. These standards were also derived from WHO figures, and it would presumably be now more difficult to reject planning policies making explicit reference to them.

Policies contained in British land-use plans create presumptions but not development rights, and the sharp end of the system is the development control stage where decisions on individual planning applications are made. In this context the relationship between planning powers and pollution control is somewhat more complex. According to the typically elusive terminology of the legislation, planning authorities must, in considering a planning application, have regard to the provisions of the Development Plan, 'and any other material considerations' [6]. The courts have been reasonably flexible in determining the boundaries between materiality and immateriality, and there is little doubt that the pollution implications of a proposed development are legitimate considerations.

The use of planning conditions, however, as a means of pollution management is rather less sound, partly on the grounds of their appropriateness as an instrument of control and partly because of the doctrine of exclusive jurisdiction. Planning decisions can conceptually be regarded as decisions to land land-based rights — subsequent alteration of, say, planning conditions involves the taking away of rights and payment of compensation. The system is therefore potentially powerful as a mechanism for anticipation, but too crude to be used as a management tool in respect of operational development. Effective pollution control must allow for flexible management, and indeed most specialized pollution consent systems incorporate powers to change consent conditions without compensation. The consent holder may be granted some protection against arbitrary or excessively costly changes by the provision of the right to appeal or the requirement of a minimum period during which conditions may not be altered without payment of compensation [7].

Even where planning conditions may appear to be an appropriate mechanism of pollution control, there remains the question of whether the existence of an alternative, more specialized control system precludes the use of planning conditions which might duplicate or run counter to such controls. The concept of exclusive jurisdiction, which is most likely to be raised if the planning authority and control agency are distinct bodies, has been little explored by the courts, though case-law indicates that the availability of alternative powers does not *per se* render planning conditions performing their functions invalid [8]. According to recent authority, planning conditions must be imposed [9],

> for a planning purpose and not for any ulterior one, and ... they must fairly and reasonably relate to the development permitted. Also they must not be so unreasonable that no reasonable planning authority could have imposed them.

This approach allows considerable latitidue, though the stance of the administrator is more likely to seek to avoid duplication of powers. Government advice upon the introduction of a licensing system under Part I of the Control of Pollution Act 1974 exemplified this approach:

> the introduction of the licensing system represents a major change in the application of planning legislation to waste disposal facilities. In the absence of specific controls, conditions attached to a planning permission have in the past been used for the purpose of water protection and public health which are now met by the licensing system [10].

The legislation concerning waste disposal on land is, in fact, the only example of pollution laws which makes explicit reference to land-use planning controls and expressly requires planning permission to be in force prior to any granting of a site licence for waste disposal. Even then there remain a number of areas where the delineation between subject matter appropriate for planning conditions and that more suitable for site licence conditions is not precise. In other areas of pollution control where pollution laws provide little in the way of anticipatory licence-type powers, greater use of planning conditions can be expected; noise is a clear example [11]. It might be argued that the extension and development of more specialized pollution controls (such as the waste disposal licensing system, or the prospective air pollution controls for local authorities) would lead to a detachment of the land-use planning system from this area of concern. This does not, in my view, seem a likely development, especially in the light of the role that planning controls will play in the field of environmental assessment.

2.2 Environmental assessment

Environmental impact assessment has been described as a 'systematic examination of the environmental consequences of projects, policies and programmes' [12]. Such a generalized description carefully avoids prescribing the forms of such an assessment or the scope of relevant considerations, although the concept has always been associated with predictive rather than retrospective evaluation. In Britain, environmental impact assessment has been particularly linked with land-based development projects such as highways or major industrial works, though until 1988 neither the term, 'environmental impact assessment' or an equivalent was to be found in existing legislative provision. Against that background, it was largely left to the discretion of both developers and planning authorities to determine the circumstances, scope and form of any assessments that are undertaken. Statutory provisions enable local planning authorities to demand further particulars and evidential verification of planning applications and include the power to require, 'such further information as may be specified ... to enable them to determine that application' [13]; on such an unassuming legal basis rested powers in respect of environmental assessment of private development. In practice, formal assessments that have been undertaken in connection with planning applications have been often initiated by the developer voluntarily, with an impetus from American oil companies developing facilities in Scotland in the late 1960s and early 1970s, accompanied by the use of environmental assessment techniques by British companies who had gained experience of them from their operations in the United States. Experience to date has been distinguished by a largely non-prescriptive approach to the subject and the varied nature of assessment forms adopted.

2.3 The EEC directive [14]

The Commission of the European Community began a development of Community environmental policies in 1972 as a result of the Paris Declaration — the same year that Britain joined the EEC. Policies emphasizing an anticipatory rather than a purely reactive approach towards pollution issues were developed, and environmental assessment was seen as a major element in such an approach. The fairly nebulous nature of the British planning control system does not encourage its role as an international legal model, and it was probably hardly surprising that the provisions of the US National Environmental Policy Act 1969 which on the surface appear sharper and more purposefully directed provided at least an inspirational starting point if not a final template. Work on the subject began in 1975, with initial drafting of Directives taking place in 1977 and 1978; following a major symposium in 1979, the first proposal for a final draft Directive was formally published in 1980. This was already the twentieth version, emphasizing the difficulties of incorporating a complex decision-making procedure within existing legal and administrative traditions of all member countries of the Community.

During the next four years, the Draft was subject to intense discussion and criticism with official responses from the European Parliament (largely favourable) [15], and, within the United Kingdom, from the House of Lords Select Committee on the European Communities [16] (again favourable) and the House of Commons (less so) [17]. During this period, the measure was consistently opposed by the UK Government principally on the grounds that the existing planning controls rendered the requirements otiose within the United Kingdom and that the particularity of the EEC proposals were incompatible and potentially damaging to current practice. Discussions then moved to the Council of Ministers, the final legislative body with the European Community, where they remained largely behind closed doors. During 1984, however, the British Government announced that the draft Directive as negotiated during this period was largely acceptable to the UK, and with the withdrawal of the principal objectors, agreement was reached in 1985, with the Directive requiring implementation in 1988.

NEPA may have provided impetus for the proposed Directive but few of its key characteristics are to be found in text of the Directive. In particular: (i) The scope of the Directive is restricted to land-based development projects, and does not encompass plans and policy proposals of a type that have regularly been subject to NEPA requirements. The reason for such a limitation appears to rest on any firm principle concerning the proper remit of environmental assessment but is largely due to the practicalities of securing agreement to the proposal throughout the Community. (ii) The term 'environmental impact statement', after making a single appearance in an early consultants' report, is never used explicitly. The concept of environmental assessment developed by the European Community shies away from the production of authoritative and possibly encyclopaedic written assessments as forming a point of focus for the procedure. Instead, there is emphasis on the *process* of decision-making, a process that largely rests on a dialogue between the developers and the 'competent authority' responsible for the final authorization (normally a local planning authority in the UK context), with, at certain stages the involvement of other parties, including members of the public, other specialist agencies, and, in certain cases, authorities in other countries where transboundary effects are envisaged. Assessment implies the evaluation of factual information, but it can be seen from the regulations implementing the Directive in Britain [18] that although developers may be obliged to produce 'environmental statements', these are envisaged as largely descriptive, data-bases rather than elaborate assessments. The planning authority is under no legal duty to justify its decision with its own written assessment, but must merely take the information about environmental affects gathered during the process 'into consideration' before reaching its decision. Though

British courts, at least, are generally less interventionist than their American counterparts, the spectre of litigation and 'second looking' of assessments have undoubtedly encouraged the diminution of specific obligations to produce written material. (iii) The EEC proposal incoporates explicit lists of development types that should be subject to the process. This aspect of the proposal has probably caused most concern to British negotiators since, though the use of broad class lists as a means of delineation or limitation is not unknown in British planning laws, their use in this context appeared to run counter to a prevailing national philosophy that local environmental considerations (which cannot be incorporated into a list class) are a critical determinant in assessing the likely impact of development proposals. Similarly — though probably for different reasons — the key provisions of NEPA leave open which types of Federal action bring into play the provisions, and it has been left to both courts and administrators to determine the practical meaning of these critical terms. In contrast, the ECC Directive largely removes this area of discretion by specifying classes of development to which the procedure must apply in all cases (Annex I list) and those to which it may apply (Annex II list), subject to criteria and thresholds to be developed by Member States.

Several explanations can be found for central importance given to the listing system (which failed to appear only in the initial consultants' report on the subject in 1976 and which was, significantly perhaps, written by two British experts) [19], even though from a technical point of view it appears to be less than perfect. The powerful precedent of French legislation, the Nature Protection Act of 1976, which was the first European legislation to make explicit reference to environmental impact assessments (*études d'impacts*), was a significant factor since this employed the listings approach. A policy presumption in favour of securing an equality of burden throughout the Community together with the need to ensure visible compliance with the new requirements throughout the member states further encouraged the use of lists. Finally, the use of listings can be expected to appeal to the administrator faced with novel obligations since their likely impact and cost can be more precisely determined.

Certainly, the implementation of the EEC Directive in its present form within the United Kingdom will have a far less dramatic and immediate effect than did the NEPA requirements on Federal decision-making within the United States. Nearly all of the development types to which the procedure would of necessity apply are already subject to close scrutiny within the existing planning control, and in this respect the Directive can be seen to have a neutral effect on current practice. More subtle effects are likely though. It is arguable that the listing approach will imply that the non-inclusion of a particular development type will justify a less rigorous scrutiny than might have occurred under existing procedures, even though significant local environmental impact is predicted. Given current pressures on local authority resources, this must be a distinct possibility. But an equally possible effect of the adoption of the Directive — and one which I feel is rather more likely — is that the obligatory use of the procedures in the context of a small number of proposals will have a spill-over effect, and will affect the approach of both developers and planners across a range of project types. The information requirements specified in the Directive may act as a model form of specification to be modified as appropriate for any proposal considered to have significant implications, even if outside the Lists.

Central to the scope of environmental assessment in its European format is the anticipation of pollution loads and problems arising from industrial developments. Since it is the existing town and country planning system which will be the chief vehicle for the implementation of the Directive within the United Kingdom, a lasting effect of the proposal will be to forge explicit links of a legal nature between development control powers and pollution management.

3. INTEGRATION

3.1 Need for an integrated approach

The development of British pollution laws has followed a pattern which must be common to many jurisdictions. Laws were developed in a fairly haphazard manner, and generally concentrate on providing controls by treating each medium — air, water and land — as a distinct area, with responsibility for enforcement resting with various specialized agencies, on a central or local basis, again acting within each medium on a discrete basis. This largely remains the picture today with principal responsibilities divided between Regional Water Authorities (inland and coastal water pollution), County Councils and other local authorities (waste disposal on land), District and London Borough Councils (noise), and with responsibility in the field of air pollution divided between the latter authorities and the Department of the Environment.

These institutional and legal arrangements have in recent years been criticized as inadequately reflecting the need to adopt control policies which fully take into account cross-medium effects. The term 'best practicable environmental option' was formulated by the Royal Commission on Environmental Pollution in its 5th Report as an appropriate criterion to apply to pollution policies [20], and while it is a concept that has been endorsed by Central Government, it has proved more difficult to agree upon appropriate institutional arrangements that can ensure that the BPEO is most effectively secured.

3.2 BPEO in practice

The appropriate response in legal and institutional terms appears to depend critically upon a sound technological assessment of the true extent and nature of cross-medium effects, including the level of decision-making at which they are most likely to occur; but there is as yet little in the way of comprehensive data or analysis on this subject. Examples of current pollution problems and the possible effects of particular strategies on pollution loads in other media may be readily given; for example, the waste disposal problem created by the adoption of fluidized bed combustion processes in power stations, or the land disposal problems created by the proposed elimination of titanium dioxide into estuarine waters. Even the alleged transfer of pollution loads to another country as a result of tighter controls adopted nationally has been presented as a failure to adopt the best practicable environmental option [21].

If one had a clean slate in establishing administrative arrangements — a luxury rarely afforded to policy-makers — an extreme approach would be to replace all existing agencies with integrated environmental control authorities. Such a response, however, ignores justifications for the present disparate nature of enforcement agencies which can to a degree be traced to the nature of the problem with which they are dealing as much as by purely historical traditions. Water Authorities, for example, are based upon principal river catchment areas rather than historical administrative boundaries; the effects of noise pollution, in contrast, are principally restricted to the immediate locality of the source, and control is therefore placed within the 'bottom' tier of local government. Other solutions proposed as administrative mechanisms to ensure consideration of the best practicable environmental option include the creation of a Central Government Inspectorate covering the whole pollution field (advocated by the Royal Commission on Environmental Pollution) to the establishment of a high level advisory body providing advice on the subject at a purely strategic level (suggested by the Confederation of British Industry) [22]. Current Government response is more circumspect. Under various pollution laws there are a number of requirements on agencies to consult with other bodies before making decisions, but there is no particular consistency, and in the main the preferred approach to date has been the encouragement of cooperative policies and arrangements between control agencies. Certainly, it could be argued that the most appropriate stage under existing procedures for the analysis and assessment of the trade-offs between the different control strategies must be when a planning application for a proposed development is considered — it is then that the implications of a development are considered in the round. The introduction of an environmental assessment procedure within the planning system may, as argued above, reinforce this process. Reliance on planning procedures, though, is unlikely to be sufficient in itself, since such procedures come into play only when new development as defined in planning legislation is proposed. Upgrading of equipment within existing plants, for example, may well have pollution implications but is unlikely to involve planning procedures. Practical experiments involving local pollution panels, involving representatives from local industry and relevant control authorities, have already been carried out and may have a positive and extended role to play in the search for effective procedures. My own feeling is that as yet there is insufficient knowledge about the extent and nature of the problems posed by the BPEO concept and of the opportunities lost by existing arrangements to justify immediate major upheavals. There is, however, a strong argument for detailed analysis of the problem, and a case for combining existing specialized pollution agencies within Central Government; this would not in my judgment increase bureaucratic load and would add as a positive stimulus to the analysis of cross-media effects and solutions. This appears to be the approach currently being pursued by the Government. In 1987, a number of the former Inspectorates were combined to form a single body, Her Majesty's Inspectorate of Pollution. Its legal powers of direct control are largely confined to air pollution and radioactive waste, but it has recently been proposed that it should have clear regulatory powers over discharges to all media from certain classes of processes or industries [23].

4. PARTICIPATION

4.1 The concept of participation

Pollution management which rests largely on processes of regulation inevitably results in the development of quite complex relationships between enforcement agencies and those parties that they are empowered to regulate. This is particularly so in the United Kingdom where the more familiar implications of enforcement discretion have been compounded by the considerable lattitude given to authorities to determine permitted levels of pollution. Other parties may claim an interest in the process of regulation, their legitimacy being based on, perhaps, the possession of threatened legal interests, such as fishing rights. The nature of third party claims may be extended by considering the external environment to be a property in which there are common and public interests. Alternatively, from a purely administrative perspective, it can be argued that public involvement will result in a decision that is sounder in the long run, because of greater understanding in its implications or perhaps because third parties will have raised important issues which might otherwise have been ignored — in this respect public participation may have an important role in raising subject-matter relevant to the search for the best practicable environmental option.

Demands for public involvement in decision-making are not a new phenomenon, but will play an increasingly important part in the regulation of pollution. However, it is important to recognize that the subject of participation raises not only the question of who is to participate but exactly what is meant by 'participation'. In the context of pollution control I would argue that it implies three broad areas of rights: information, involvement, and enforcement. Each area may involve a considerable range of possible rights, and while there are obvious strong linkages between the three areas, the different classes of rights are not necessarily mutually dependent on each other [24].

4.2 Information

Of the three classes of right, that concerning access to information concerning pollution loads and individual consent standards raises perhaps the least fundamental questions, despite it being an issue of current controversy. This is not to deny the practical difficulties of making available such information but modern information technology undoubtedly removes much of the administrative costs associated with public access, and a number of enforcement agencies have already moved a considerable way towards the provision of information on a more general basis without the need for legislative backing.

There are, however, two important areas where more extensive research of a technical nature would assist the designer of legislation. First, there appears to be little detailed information on the relationship between information concerning pollution discharges and the prejudicing of commercial or trade secrets. I am prepared to accept that this may be a subject area where it is not possible to arrive at useful conclusions, and that legislation must simply allow for the resolution of a possible conflict of interests to be determined on a case by case basis (the approach adopted in Part II of the Control of Pollution Act 1974) [25]. This, though, will not absolve the legislator from making presumptions as to the grounds of exemptions from disclosure of information, if exemptions be allowed, and as to the party upon whom the burden of proof must rest. The second critical area concerns the likelihood of misunderstanding of publicly available information of a technical nature, especially when the underlying philosophy of pollution control is largely based not on the absolute prevention of discharges but on the use of the natural environment as a pathway for disposal and dispersal. The possibility of public misinterpretation of information does not, in may view, warrant justification for its withholding, but it is probably complacent to assume that greater sophistication amongst emenity groups and other interested bodies will provide a solution to the problem. The form in which data is presented including the nature of accompanying interpretative statements and the impact this has on public understanding is likely to demand considerable analysis from a range of disciplines in the future.

4.3 Involvement

The direct involvement of third parties in regulatory decision-making has never been as extensive as that found in the development control system, where there has long been a strong tradition of promoting at least local public involvement. It is, of course, rather more difficult to incorporate participatory procedures into a regulatory process that by its very nature is likely to be diffuse and iterative, in contrast to what are essentially 'one-off' decisions concerning planning application; the irreversible and singular nature of planning decisions no doubt account for the focus that they currently provide for public involvement. The major exception to this 'closed system' of regulation is now the procedure for granting consents for trade or sewage effluent discharge into waters, with the innovative provisions concerning public participation introduced in Part II of the Control of Pollution Act 1974 being implemented by 1984 [26].

Forms of participatory procedures are varied, ranging from a simple right to make written representations to the more extensive types of public examination of which the local public inquiry is the most well-known example. Aside from the choice of procedure to be adopted, the introduction of participation rights necessitates the formulation of suitable screening criteria which will filter out cases of little significance. Legislation such as the Water Resources Act 1963 governing abstraction licences adopts extensive classes of exemption contained in the principal statutory provisions, whilst the Town and Country Planning Acts grants discretion to Central Government to determine the extent of participation permitted beyond the right to examine public registers of planning applications. In contrast to these mechanisms, the provisions in Part II of the Control of Pollution Act 1974 raise a presumption of public participation in all application decisions, but allow Water Authorities to exclude such participation where they judged that a proposed discharge 'will have no appreciable effect' on the waters concerned [27]. This approach, which is consistent with the belief that environmental impact is critically affected by local considerations, demands public confidence in the ability of the relevant authority to exercise its judgement and discretion with good sense, and the approach adopted by Water Authorities or their successors [28] towards their task in future years — and the public reaction this may generate — will be worthy of study.

4.4 Enforcement

In the absence of express statutory restrictions, English law has long permitted any citizen to initiate private prosecutions in respect of an alleged criminal offence, whatever the nature of their interest in the matter. Protective devices exist to prevent vexatious or totally groundless litigation, but in the field of pollution control there are many examples of express statutory prohibitions on prosecutions being brought by those other than public officials. The extent of discretion afforded to public agencies in the legal prosecution of non-compliance with pollution controls permits a wide range of approaches across different agencies, even when they act within the same field. Part II of the Control of Pollution Act 1974 provides an unusual recent example in recent years of the restrictions on prosecution being removed, and it is therefore an area which may clearly demonstrate the effect of the availability of the right. Because of the delayed implementation of the provisions the first private prosecution did not take place until 1987 [29], but already the possibility of this has had a strong influence on the design of individual consents. First, standards expressed in consents can now be expected to express realistic compliance levels rather than optimistic targets which had been a characteristic of previous consent standards. Secondly, the compliance rate which is determined for such consents no longer rests within the remit of administrative decision but, in respect of discharges by Water Authorities at least, have been translated into legal requirements as formal consent conditions [30].

The decision to permit private prosecution in the field of pollution control may not alter the levels of pollution permitted or reached in practice. But the availability of the right can be expected to demand greater clarity in control policy together with a restriction on administrative freedom of action as legal formalism replaces prior discretion.

4.5 Explicitness

Much of British pollution law has hitherto remained fairly inexplicit in any statement of precise aims. Detailed frameworks of decision-making are expressed, but the objects of controls may be unstated (as in the case of the legislation governing discharges of trade and sewage effluent) or may be expressed in generalized and non-quantitative terms (for example, the duty to supply 'wholesome' water to domestic consumers).

This characteristic of British pollution law is currently under considerable challenge by new policy demands, a trend that is unlikely to discontinue. Increased opportunities for public participation in the process of regulation, in whatever form such opportunities take place, are likely to require more explicit explanations of policy approaches, and possible restructuring of consent or licence conditions. But more significant influences on the level of explicitness to be found in legislative measures result from both the continued elaboration of EEC policies in the environmental field and the further development of the Governments privatization programme, particularly of the water and electricity industries. This is not to rehearse the old arguments concerning the conflict between the British and Community approach towards pollution control but reflects more fundamental attributes of EEC legislation. The establishment of Community institutions with responsibility for the monitoring and enforcement of Community obligations is a feature unknown in other international arrangements where responsibility for compliance is generally left to the discretion of subscribing parties. This provides a sharpness to Community legislation, and encourages the expression of standards or goals in numerate rather than purely qualitative terms, with a consequent effect on national legislation and practice; the term 'wholesome' mentioned above is undefined in British law but has in effect been translated into quality standards expressed to numerate limits as a result of EEC legislation on water quality [31]. EEC environmental legislation has to date generally been expressed in Directives rather than Regulations, giving individual member states considerable discretion in determining the measures needed to secure compliance with the obligations imposed upon then by such Directives. But such discretion may not be all that it seems, and recent case-law in the European Courts suggests that member states do not have total freedom in the manner of compliance:

> Mere administrative practices which by their nature can be changed as and when the authorities please and which are not publicized widely enough cannot in these circumstances be regarded as proper fulfillment of the obligation imposed by Article 189 on Member States to which the Directrives are addressed.
>
> EC Commission v Belgium (1982) [32]

The doctrine of open compliance is still being developed by the European Court and its full implications for countries such as Britain which has often favoured discretionary administrative practices as a means of meeting Community obligations have yet to be felt [33].

This move towards a more transparent and formal system of pollution regulation has been reinforced in a national context by current government policies favouring privatization of public sector industries. The regulation of public corporations, whose operational activities can be presented as being in the 'public interest', buttressed by explicit statutory duties and sometimes overseen by a sponsoring, non-environmental governmental department, raises a distinctive set of administrative and institutional tensions [34]. Reconciling potential conflicts of interest are made all the more problematical where, as in the case of sewage treatment works operated by water authorities and local authority waste disposal sites, the agency charged with the responsibility of pollution regulation is at the same time involved in potentially polluting activities. The decoupling of production and regulatory functions which is now a core element of current plans to privatize the water industry and may be expected to extend to other areas, does not inevitably have to lead to a more formalized system of pollution regulation, but at the very least will encourage and assist the trend.

It is clear that in many areas of British pollution policy, there have been recent developments, of both international and national origin, which are set to change well-established norms of law and practice. A sea-change in the process and style of formal legislation is occurring. The emphasis on the need for soundly based and practical solutions to pollution problems, coupled with a large element of administrative discretion, which has long coloured practice in this field will remain powerful. But increasingly policy-makers and those responsible for law enforcement will be faced with reconciling this approach with other, potentially competing political and social demands of the type described in this chapter. The extent to which these elements can be married to reinforce rather than inhibit appropriate technical solution to pollution issues will be a major challenge for the next decade.

REFERENCES

[1] Heap, Sir Desmond (1982) *Outline of Planning Law,* Sweet and Maxwell, London.
[2] Jowell, J. and Noble, D. (1981) Structure plans as instruments of social and economic policy, *Journal of Planning and Environment Law,* 466–480; Loughlin, M. (1980), The scope and importance of 'material considerations', *Urban Law and Policy* (1980), 171–192.
[3] Town and Country Planning Act 1971, s 7 (3) (a).
[4] Miller, C. and Wood, C. (1983) Air pollution in Cheshire, in *Planning and Pollution,* Clarendon Press, Oxford.
[5] Directive on Air Quality Limit Values and Guide Values for Sulphur Dioxide and Suspended Particulates, OJ L229 30.8.80.
[6] TCPA 1971, s 29 (1).
[7] For example, see Control of Pollution Act 1974, s 38.
[8] See the decision of the House of Lords in Westminster Bank *v* Beverley Borough Council [1971] AC 508.
[9] Per Viscount Dilhorne in Newbury District Council *v* Secretary of State for the Environment [1981] AC 578.
[10] DOE Circular 55/76, para 43, HMSO, London, 1976.
[11] See DOE Circular 10/73, *Planning and Noise,* HMSO, London 1973.
[12] Clark, B. (1980) *Environmental Impact Assessment,* Mansell, London.
[13] Town and Country Planning Act 1971, s 25.
[14] Directive concerning the Assessment of the Environmental Effects of Certain Public and Private Projects (EEC/85/337). See generally Haigh, N. (1983) The EEC Directive on Environmental Assessment of Development Projects, *Journal of Planning and Environment Law,* 585–596 and Sheate, W. (1984) *The EEC Draft Directive on the Environmental Assessment of Projects — its History, Development and Implications* (unpublished MSc thesis), Imperial College, London.
[15] *Official Journal of the European Community 1982,* C66.
[16] *House of Lords Select Committee on the European Communities, 11th Report Session 1980–81,* HMSO, London.
[17] House of Commons, *First Standing Committee on European Community Documents,* 9 June 1981, HMSO, London.
[18] Town and Country Planning (Assessment of Environmental Effects) Regulations 1988 S.I. 1988/1199 covering most land-use projects with the Directive. Similar parallel regulations have been made concerning certain other projects, such as highways.
[19] Lee, N. and Wood, C. (1976) *The Introduction of Environmental Impact Statements in the European Communities,* CEC, ENV/197/76, May 1976.
[20] Royal Commission on Environmental Pollution (1976) *Air Pollution Control:*

an Integrated Approach, 5th Report, Cmnd. 6371, HMSO, London. See further the Commission's most recent report, 'Best Practicable Environmental Option', 12th Report, Cmnd. 310, HMSO, London.

[21] ENDS Report (1985) *Pollution Transfer (concerning rising imports of asbestos cement),* Vol. 121, February 1985.

[22] Royal Commission on Environmental Pollution (1984) *Tackling Pollution — Experience and Prospects,* 10th Report, Cmnd. 9149, HMSO, London.

[23] UK Department of the Environment, *Integrated Pollution Control,* A Consultation Paper, DOE, London, 1988.

[24] These ideas are further explored in Macrory, R. (1985) *The Role of Third Parties in the Regulatory Process* — paper to International Bar Association Seminar, Industry and the Regulatory Agencies, Stratford, England, April 1985.

[25] Control of Pollution Act 1974, s 42.

[26] See DOE Circular 17/84, *Water and the Environment: The Implementation of Part II of the Control of Pollution Act 1974,* HMSO, London, 1984.

[27] Control of Pollution Act 1974, section 36 (4).

[28] Under the Water Bill 1989, it is proposed to privatize the ten existing regional water authorities in England and Wales, and place their pollution control functions in the hands of a new single national agency, the National Rivers Authority.

[29] See Environmental Law, *Journal of U.K. Environmental Law Association,* Vol. 2, Nos 1 and 2.

[30] Department of the Environment (1985) *Statement on Policies and Procedures for the Control of Water Authority Effluent Discharges,* 30 January 1985.

[31] EEC Directive Relating to the Quality of Water Intended for Human Consumption, 15 July 1980, OJ L 229, 30 August 1980.

[32] 1982 Common Market Law Reports 627; and see Haigh, N. (1984) *EEC Environmental Policy and Britain,* Environmental Data Services, London.

[33] See further, Macrory, R. *Industrial Air Pollution: Implementating the European Framework,* Proceedings of the National Society of Clean Air Annual Conference, Llandudno, 1988. For a general analysis of EEC environmental policy and its effect in Britain, see Haigh, N., *'EEC Environmental Policy and Britain',* 2nd edn, Longmans, London, 1987.

[34] The issue is further discussed in Macrory, R., Regulating state enterprises — a comparative legal model, in *Yearbook of Soviet Law* (ed. Butler, W.), 1989, forthcoming.

6.3

Evolution of technology strategies in environmental policy formulation

Professor David H. Marks and **Deborah L. Thurston**
Massachusetts Institute of Technology, Cambridge, Massachusetts

1. INTRODUCTION

Early analysis of 'the environmental problem' by economists has been proven essentially correct. Environmental pollution results when economically rational waste dischargers do not include the externality of the harmful impact of their activities on the environment when making decisions. Only those costs and benefits which accrue directly to the decision maker are considered.

The economists' solution to the problem has been simply to internalize the externalities, or put a price on pollution. This approach was partially reflected in the requirements of the Clean Water Act Amendments of 1977 for wastewater treatment systems, whose cost would accrue to the discharger. The Amendments incorporated technology based standards, designed to advance the state of the art of wastewater treatment technology. Increasingly stringent waste treatment standards have provided the incentive to industry to develop more efficient and less costly waste treatment technology, resulting in higher levels of pollutant removal prior to discharge.

In developing more efficient advanced waste treatment systems and complying with these regulations, industry has for the most part demonstrated a tremendous ability to adapt and respond to economic incentives. Our goal now should be to capture this flexibility of industry when formulating water and wastewater policy in order to provide incentive which will direct industrial technolgical change in the most efficient direction.

Making basic changes in the water using and wastewater generating process itself in order to decrease the amount of waste produced frequently also results in increased efficiency in other areas as well, including reduced raw materials, labour and energy requirements. When compared with 'end-of-pipe' waste treatment systems, which incur cost to the discharger with no benefit other than compliance with discharge standards, process design change is frequently the more economically efficient alternative.

Given the response of industry to previous technology forcing policy, the time has come for an evolution, or a change in focus of that policy. The focus should be switched to, or at least broadened to include the water using systems responsible for wastewater production, rather than solely the end of the discharge pipe.

This chapter will discuss causes of the counterproductivity of the existing policy and propose a direction in which policy may evolve to result in more efficient systems.

2. TRADITIONAL APPROACHES TO WATER SUPPLY AND TREATMENT

Water supply management has traditionally concerned itself with demand forecasting and planning to meet that demand in an optimal fashion. Demand has been seen as a predetermined inelastic quantity which must be estimated, rather than a variable which may be influenced and changed. A clean, unlimited, low cost water supply is often viewed as a basic, inalienable right in developed countries. Restrictions on consumption or use are rare, and usually instituted only in cases of drought or other water supply emergency [1].

However, mechanisms such as metering or rationing used to decrease demand during emergency situations have been successful, which indicates that demand is indeed elastic [2]. Normal demand may also be reduced via conservation devices such as shower flow controls [3]. It has also been shown that large water rate increases reduce domestic water demand [4].

In spite of this, demand is often assumed to be inelastic for planning purposes, and the price of water is often set so low that demand is indeed inelastic. Small price increases do not result in decreases in domestic (inside) demand [5], since even the new, higher price is still insignificant in relation to the household budget. The pricing system therefore does not provide economic incentive for conservation. Some pricing systems, in fact, provide economic disincentive for water conservation, such as decreasing block rate structures common for industrial users.

In contrast, wastewater discharge in the United States is strictly regulated. Specific limits are defined as to the quantity, quality and location of discharge via the National Pollutant Discharge Elimination System (NPDES) permit. The required level and type of wastewater treatment technology is frequently specified, as in 'secondary standards' for municipal systems and categorical pretreatment standards for industry such as 'Best Practicable Technology' (BPT).

This system has given industrial and municipal users the message that they will be held financially responsible for their water using activities via wastewater treatment costs, but not through water supply costs. Even when water price has been recognized as influencing demand, its relation to wastewater generation and subsequent treatment requirements has been ignored [6].

3. RESPONSE OF WATER USERS AND WASTEWATER GENERATORS

Continuing low water prices coupled with increasingly stringent wastewater treatment standards have provided market incentives to dischargers to develop less costly, more advanced waste treatment technology rather than to decrease the overall quantity and/or level of harmful constituents entering the waste stream. Treatment technology has been greatly advanced, making greater and greater levels of treatment technically and economically feasible.

Many industrial processes continue to generate large quantities of high strength waste and are by definition 'wasteful' or inefficient, although system managers may be behaving perfectly rationally given the low price of water and the now low cost of treating it prior to discharge.

However, given sufficient incentive, industry has shown that it is capable of responding to supply costs or discharge standards with process change resulting in decreased demand for water and wastewater treatment in addition to or in lieu of treatment technology advances. Examples are:

(a) The price of water was found to be a significant factor in variations in demand for the fruit and vegetable preservation industry [7].
(b) When the demand for treatment is viewed as analogous to the demand for energy, the long-term response to the 'energy crisis' is an example. The 'crisis' resulted in elimination of many energy wasteful practices and processes, with much research devoted to but few significant increases in the use of alternative energy sources which would cheaply satisfy ever increasing demand.
(c) Industrial response to wastewater pretreatment regulations has in some cases seen pretreatment cases demand decrease rather than treatment technology advance; electroplaters have significantly decreased the amount and strength of waste that their process produces. This has been accomplished through use of mobile tanks for concentrated acid collection, substitution of less harmful chemicals for more toxic ones, separating and recycling cooling water, use of oil skimmers, etc.
(d) A survey of 30 large plants suggests that firms do adjust demand in response to even small price changes, in addition to technological improvements [8].

4. CONTINUATION OF EXISTING POLICY IS COUNTERPRODUCTIVE

It is inefficient to continue in the original direction of forcing continuing treatment technology advances in the absence of incentives to decrease demand for that treatment. Technology forcing regulations are those whose purpose is to bring about a technological change in such areas as process design or treatment methods in order to lower treatment cost and/or decrease the amount of pollutants being discharged to the environment. Such regulations typically set a compliance deadline some months or years in the future for meeting new discharge standards or paying discharge fees that are not in force at the time the legislation is drafted. Thus, in order to comply with the new standards or fees in an economically efficient manner, the polluter must do one of three things:

1. Employ treatment technology that currently exists but is not in use. This would include 'conventional' treatment systems, such as trickling filters for wastewater treatment plants.
2. Develop and employ new waste treatment systems, or adapt a treatment techno-

logy used in a different industry. An example would be treatment of toxic waste by enzymatic detoxification.
3. Make basic changes in the process design (including materials substitution) that is responsible for the production of pollutants, so that pollutant formation is decreased or eliminated entirely.

Failure of a technology forcing policy can result from too lax a discharge standard or fee, which may encourage add-on or 'band-aid' treatment systems, rather than basic design changes. This can occur when the cost of the add-on devices is relatively low when compared with changing the process responsible for pollutant production. Add-on solutions can also be the result of a deadline that is too stringent, although the standard or fee itself would induce process change if adequate compliance time were granted. Insufficient compliance time would force industry to expend time and money on 'end-of-pipe' treatment solutions, which could be arrived at more quickly than changes in process design. Therefore, if the desired goal is basic process design change, rather than add-on treatment devices, the discharge standards must be set out of reach of existing low cost treatment technologies, and the deadline for compliance must allow enough time for research, development and planning necessary to institute process changes.

The problem with the existing system is that treatment technology has been forced or advanced to the point where it is in some cases less expensive than the cost of process design change which would remove the same amount of pollutants from the waste stream. While it is in general a definite improvement that treatment costs have been lowered, they have in some cases been pushed so low as to discourage process change. Treatment systems generally impart high capital, operating and maintenance costs, with no corresponding benefit to the discharger other than compliance with regulations. Process change, on the other hand, frequently benefits industry in other aspects, such as lower materials, energy and labour costs.

The basic framework of the popular cost benefit and cost effectiveness analyses approaches also increases the likelihood of inefficient systems. Such analyses consider the environmental impact or 'benefit' of various levels of water quality, and weigh those against the cost of varying levels of wastewater treatment necessary to achieve those levels of water quality. When the cost of treatment just balances the benefit of water quality, the proposed system is said to be cost effective. These analyses assume that a given amount of wastewater is generated, with constituents being either removed through treatment or discharged to the receiving water. They generally do not consider the possibility of changes in the amount of waste generated within the planning area. In reality, it may be much less costly to eliminate certain particularly harmful constituents or decrease the amount of waste produced than to treat the original waste stream to the 'cost effective' level. It may be better to spend the next dollar on conservation measures rather than more advanced treatment technology, particularly since that technology experiences increasing marginal cost at more and more advanced treatment levels. It is much more expensive to remove the last 5% of pollutant load than to remove the first 5%.

Similarly, after the desired treatment level is defined, for example 'secondary treatment', preliminary design and cost estimation is carried out to determine the most cost effective alternative treatment system for meeting the standard. Such a method assumes a given standard of treatment in addition to a constant waste stream.

Using technology based discharge standards to continue to force the development of more advanced waste treatment technologies may be more costly both to dischargers and to the environment than encouraging process change. If the rate of treatment technology advance and the correspondingly more stringent technology based discharge standards do not equal or exceed the rate of industrial growth, an ever-increasing total waste load on the environment could result. However, lesser amounts of waste generated, even if treated to less stringent standards, may result in lower treatment costs and lower overall discharge of pollutants.

5. PROPOSED SOLUTION

Rather than continuing to force technology only towards the direction of greater and greater levels of advanced wastewater treatment technology, we should also force technology from the water use and waste production side. It should be possible to combine the municipal management tasks of water supply and wastewater treatment, viewing the water use, contamination and discharge system as the cycle that it is, rather than dissecting it into unconnected parts. The basic idea is to provide incentive for process change resulting in conservation from *both* the water supply cost and treatment cost directions. We should encourage waste generators in discovering and incorporating water use practices and process design change which improves efficiency by using less water and/or by producing less polluted waste, in terms of both quantity and quality.

Economic incentive for water conserving process design change could be provided in the form of increased water prices, into the range in which elasticity of demand is observed.

Theoretically, rates could be set based on a thorough technical understanding of the process responsible for generation of the waste stream. Price could be set to promote process change resulting in decreased water use and discharge. This could be accomplished by setting the marginal cost of water to each user just greater than the marginal cost of conservation measures available to that user. Increasing block rate structures could be used to account for the increasing marginal cost of greater levels of water conservation and decreasing marginal cost (or economy of scale) of treatment.

However, such a fine-tuned model would require significant study of the water using processes of each user in the planning area. In order to avoid studying each water using system to death, it would be sufficient at the start merely to increase water prices to just past the point at which elasticity of demand is beginning to be observed, and imposing a general trend towards continuing, gradually increasing water prices, corresponding to the continual, gradual increase in treatment requirements that have been imposed. The point at which elasticity appears is that where the marginal cost of water just equals the marginal cost of going without that extra unit. The cost of going without is the cost of making process or materials changes necessary to get by on one less unit of water minus the cost of treating that unit of flow. While the existing system provides incentive for investment in process design change only up to the point where it is less than or equal to the cost of treatment, the proposed system will provide incentive for more extensive process design change whose cost is less than or equal to treatment cost *plus* water supply cost. The user is in the best position to determine the most cost-effective allocation of its resources between water supply costs, investment in process design change and wastewater treatment systems.

6. SUMMARY

We have used wastewater treatment technology based discharge standards successfully as a technology forcing policy. This initial emphasis has led to the development and use of advanced wastewater treatment systems on a large scale. However, changes in water using processes and operations which decrease the amount of wastewater produced have great potential to be more economically efficient than large scale advanced treatment. Process change may also include the benefit of decreased energy, materials and labour costs. The current system of wastewater treatment technology forcing discharge standards provides incentive for process changes only up to the point where they are less expensive than treatment costs. These treatment costs have been driven very low due to the technology forcing effects of discharge regulations. The proposed system of increased water prices into the range where elasticity of demand is observed would provide incentive for water conserving process design change beyond the incentive provided by existing treatment requirements. This proposal broadens the scope of alternatives that dischargers may evaluate in arriving at the most economically efficient system by which to comply with discharge standards. This would thus result in lower amounts of water demanded, polluted, treated, and discharged to the environment.

REFERENCES

[1] Russell, C. S., Arey, D. G. and Kates, R. W. *Drought and Water Supply Resources for the Future*, Johns Hopkins Press, 1970.
[2] Bruvold, W. H. Residential response to urban drought in Central California, *Water Resources Research*, December, 1979.
[3] Sharpe, W. E. Municipal water conservation alternatives, *Water Resources Bulletin*, American Water Works Association, October, 1978.
[4] Hogarty, T. F. and Mackay, R. J. The impact of large temporary rate changes on residential water use, *Water Resources Research*, December 1975.
[5] Howe, C. W. and Linaweaver, F. P. The impact of price on residential water demand and its relation to system design and price structure, *Water Resources Research*, First Quarter, 1967.
[6] Dandy, G. C., McBean, E. A. and Hutchinson, B. G. *Water Pricing Policies as an Aid in Demand Management*, Hydrology and Water Resources Symposium, November, 1983.
[7] Herrington, P. R. *Water Use in Fruit and Vegetable Processing in the UK*, Progress Paper A.5, United Kingdom Water Resources Board, September, 1971.
[8] de Rooy, J. Price responsiveness of the industrial demand for water, *Water Resources Research*, Vol. 10 (3), 1974.

6.4

Environmental regulation formulation and administration

R. K. Jain
Chief, Environmental Division, US Army Corps of Engineers, Construction Engineering Research Laboratory (USA-CERL), Champaign, IL, USA, and
D. Robinson
Principal Investigator, USA-CERL, Champaign, IL, USA

1. INTRODUCTION

Introduction of an external force, such as a government regulation, in a free market economy like that in the United States, is likely to disrupt its efficiency. It is generally believed that competitive markets, devoid of government regulations, can produce efficient (i.e. Pareto optimal) outcomes. Environmental regulations, thus, can affect the cost and completion time of industrial processes. To some, environmental regulations have imposed excessive costs on businesses compared with the benefits they generate for society.

This chapter discusses several public policy issues concerning environmental regulations (such as the basis for promulgating environmental regulations, and public concerns and views about environmental regulations); the effects of environmental regulations on the economy (such as societal benefits derived from environmental regulations and subsequent costs imposed on industry and the effects of regulations on industrial productivity); and various approaches to environmental regulation and administration. Based on these discussions, the chapter focuses on inefficiencies and suggested remedies for the formulation and administration of regulations that achieve compliance with minimal direct and indirect costs.†

2. PUBLIC POLICY ISSUES IN FORMULATING ENVIRONMENTAL REGULATIONS

A number of public policy issues are important in discussing environmental regulations, such as the theoretical basis of promulgating environmental regulatory

† Judgements and opinions expressed in this chapter are solely those of the authors and should not be viewed as the position of any governmental agency.

requirements, as well as views of business leaders and private citizens about the importance of maintaining an environmental protection structure within government. A brief discussion of these issues should provide a clearer understanding of the reasons why industrialized countries like the United States are taking steps to establish environmental regulations to protect the health and welfare of their people.

2.1 Basis for environmental regulation
This section of the chapter reviews some of the theoretical and practical reasons given for imposing environmental regulations.

2.1.1 *Pareto optimality and competition*
From an economy-wide perspective, producing and consuming activities (such as agriculture, mining, manufacturing, etc.) appear to be unplanned, almost random, events. And yet there is order in the results. For example, how amazing it is that the right number of cartons of milk in the correct quantities end up on the shelves of local grocery stores, when the process starts with the conception of calves in places like California and Wisconsin — all accomplished without governmental direction. Truly, it was the 'genius' of Adam Smith who first saw that '... prices, the powerful signalling and incentive forces generated by private exchanges in markets, were at the core of a process which, via the decisions of many independent economic units transformed resources into products and distributed them to consumers' [1 (p. 18)].

It has long been recognised that if business enterprises are left to their own devices and if individual consumers are allowed to satifsfy their desires unhindered by external restrictions, society is the better for their actions. That is, when one or more persons can benefit from an exchange of resources or goods and services without negatively affecting any other person, then it is in the best interest of society for the exchange to be carried out. When no additional exchanges can be made, the economy has reached a situation where each individual cannot improve his/her own situation without damaging that of another. Hence, society reaches a point of 'Pareto optimality' or the point of greatest benefit to society given the existing distribution of resources. Simply put, Pareto optimality is a situation in which 'all possible gains from voluntary exchange have been exhausted' [1 (p. 20)].

As shown by Von Neumann [2], Arrow and Debreu [3], and more formally by Debreu [4], an 'ideal' market exchange economy like the one described above, will always lead to a Pareto optimum position. This interesting result, where each individual pursues his or her own private interests and the resultant solution, is a Pareto optimum situation which provides not only a basic principle for welfare economics but also forms the foundation underlying most economic justifications for environmental regulations. Quirk and Sapoknik [5] provide a good source of reviewing these and other important issues surrounding competitve markets and equilibrium processes.

2.1.2 *Market failure*
Why are environmental regulations necessary? Traditionally, government regulation has been justified to correct certain 'market failures' that impede the marketplace from functioning properly. Besides the existence of natural monopolies (such

Ch. 6.4] Environmental regulation formulation and administration 259

as public electricity and telephone utilities) and imperfect market information, 'externalities' have been used to justify government regulations.

Externalities arise when technological interactions between parties of an exchange are not reflected in the prices. Externalities can either impose costs or confer benefits on third parties to a market exchange. For example, industrial pollution where the producer is not required to pay for damages to those people injured by the pollution is a case of externality. Generally, one would find the externality over-supplied when the effects of the externality are not reflected in market prices. These are cases in which externalities do not necessarily lead to market failure. That is, if the injured parties (assuming a small number) can easily negotiate a settlement with either of the exchange parties, then there may be a Pareto optimum generated [6].

The following discussion focuses on several market failures or deviations from our 'ideal' market structure that have been suggested as justification for intervention by environmental regulation in the United States.

- Because labour and capital are scare resources, their use by industry is considered in the decision-making process for production and sales. The environment, on the other hand, has been considered a free commodity. One might consider that the environment is a resource held in 'common' for the population as a whole; i.e., no one person (or group of people) hold the property rights, so a market price is not established. The result is that the air, water, open lands, and other environmental resources are over-used. Consequently, there has been considerable environmental degradation with the attendant social costs. Simply put, some economic and social costs are not reflected in market exchange of goods and services.
- There are situations where health and safety considerations preclude the use of market incentives. For example, it would be unwise to put a dollar value on the discharge of such toxic materials as PCBs or mercury into the environment.
- Some projects involve the exploitation of energy and other natural resources at unprecedented rates. The question of temporal optimality or the proper rate at which certain exhaustible resources would be used becomes important. In such cases, a market economy is unable to properly account for such long-term economic and social costs, especially when ownership of the resources is questionable. Because market interest rates tend to exceed social rates of time preference in these situations [7 (p. 368)], the market-place will tend to encourage consumption of exhaustible resources too fast.

2.2 Concerns regarding environmental regulations

Many public administrators, engineers, industrialists, and other decision-makers recognize the need for environmental protection. They also realize the importance of economic efficiency and utility. There are, indeed, a number of justifiable concerns regarding environmental regulations. These concerns are shared by many who feel that environmental regulations can be structured to affect minimally the efficiency and productivity of the economy's private sector or interfere minimally with essential Federal programmes such as National defence, and still achieve reasonable environmental protection goals. Some of the concerns related to environmental regulations that have been identified are:

- Environmental regulations seem to be structured in ways that impose excessive costs in comparison to the benefits produced by the regulations.
- In general, environmental regulations are of the 'command-and-control' type. Consequently, in a 'free-market' economy, they tend to be ineffective and generate inefficencies in the use of resources because they fail to preserve the elements of voluntary choice.
- Procedures for administering environmental regulations lack properly structured incentives.
- The National Environmental Policy Act (NEPA) and other environmental regulations require unnecessary paperwork and cause unnecessary delays in completion schedules.
- Many environmental regulations are duplicated at different governmental levels and, at times, are incompatible with each other. As a result, they create unnecessary work and inefficiencies.
- Pollution control regulations require investments in equipment, operations, and maintenance which compete with investments in productive plant and equipment.
- New sources of pollution (e.g., water and air pollution) are often subjected to more stringent standards than existing sources. This can cause businesses to retain existing, less productive plant and equipment or delay the introduction of newly developed technology [8 (pp. 56–70].
- To avoid plant closing or lay-offs, environmental regulations are written and enforced more stringently for fast-growing industries than for slow-growing sectors, thus inhibiting an important source of national productivity growth. The electricity industry, which had an excellent record of productivity growth until the early 1970s, has been cited as an example [8 (pp. 56–70)].

2.3 The public view

In contrast with the concerns regarding the deleterious effects of environmental regulations on the efficiency and productivity of the nation's industrial sectors, the US public appears quite willing to incur the costs that are necessary to improve our environment. In fact, public opinion regarding environmental protection and environmental regulation has a profound effect on environmental policy in the US. During a 1980 survey conducted by Resources of the Future (RFF), a plurality of those surveyed felt that environmental protection is 'so important that ... continuing improvement must be made *regardless* of the cost' [9].

In spite of considerable economic problems facing the nation, one October 1981 survey, as reported by Phillip Shabecoff of the *New York Times* [10], indicates that:

- More than two out of three people questioned agreed that 'we need to maintain present environmental laws in order to preserve the environment for future generations'. Fewer than one in four thought that 'we need to relax our environmental laws in order to achieve economic growth'.
- When confronted with a specific question regarding keeping the air pollution laws 'as tough as they are now' even if 'some factories might have to close', nearly two-thirds of a sample of 1479 Americans said they would keep air pollution laws tough.

Ch. 6.4]   **Environmental regulation formulation and administration**   261

According to pollster Louis Harris's comprehensive survey of attitudes towards water pollution, an overwhelming 94% of Americans believe the Clean Water Act should be kept as it is or even made more stringent [11].

Often the question is raised as to whether the polls really indicate the willingness of individuals to pay for environmental amenities or whether the polls simply reflect a romantic interest in environmental quality. Benefits of pollution control and environmental protection depend upon the value one places on a clean environment. But, there is no easy way to determine the price of a clean environment. Willingness to pay only indirectly establishes a 'shadow price' for a clean environment. In another Harris survey [11], people were asked if they would be willing to pay $100 more per year in taxes or higher prices for clean and safe water. Incredibly 70% of those polled said that they were willing to pay the extra $100, as opposed to the 26% who were not.

Based upon surveys of public attitudes, in comparison with other economic and social goals, it appears that environmental issues are playing a significant role in the decision-making process of public and private enterprises [12 (p. 7)].

3. EFFECTS OF ENVIRONMENTAL REGULATIONS ON THE ECONOMY

Environmental regulations cause various impacts on our society. These impacts are felt in a variety of ways. First, environmental regulations generate cleaner environments. This has positive implication for the health and welfare of society. It contributes to preservation and protection of the life support system, and enhancement of long-term viablity of the environment. Second, environmental requirements usually increase the 'cost of doing business' for firms. And third, there is concern that recent growth in national productivity has been retarded because of environmental regulations. These topics will be discussed in this section of the chapter.

3.1 Benefits derived from environmental regulations

The effects of environmental degradation are varied and complex. Benefits from environmental enhancements are not normally marketable. As a result, quantifying the benefits from environmental regulations is difficult and empirical evaluations can be widely variable. Despite these impediments, Ashford and Hill [13] were able to review and summarize many studies and to provide a range of benefits that can be attributed to environmental controls. As an example, they estimated that the US population, in terms of health costs and reduced loss of work, benefits from air pollution regulations by as much as $58.1 billion per year and water pollution regulations by as much as $10.4 billion per year. Additional benefits, such as less damage to physical structures, property values, wildlife and the natural environment, crops, and the natural resources cannot be quantified easily, but are considerable.

3.2 Costs resulting from environmental regulations

By their nature, environmental regulations impose a variety of pollution control requirements on business establishments or restrict their way of doing business. Normally, the pollution control requirements represents 'costs of doing business'. Environmental regulations impose three kinds of costs to firms [14 (p. 18)].

First there are the administrative costs of government agencies. These are the costs, borne mainly be taxpayers, of operating the Federal, State and local government agencies that are responsible for implementing the environmental controls. Although it was a crude estimate, Litan and Nordhaus [14 (p. 19)] found that all regulatory programmes (not just environmental) cost slightly less than $600 million to administer in 1969 and grew to over $3 billion by 1979. After correcting for inflation, what was $670 million in 1969 increased to $2 billion in 1981 (in 1972 dollars).

Second, there are 'direct' compliance costs required by firms in order for them to meet the environmental regulatory requirements. In a recent article in the *Survey of Current Business*, Farber *et al.* [15] estimate that real pollution abatement and control expenditures (i.e. after taking into account inflationary changes since 1972) throughout the entire US economy rose continuously from 1972 to 1979 at an average annual rate of 5.6% but that they have declined every year after 1979 at an average annual rate of -2.3% (1979 to 1982). That is, starting in 1972, when the government, businesses, and consumers spent $18.4 billion (1972 dollars) for various pollution abatement and control items, expenditures increased continuously until 1979 with their highest level at $26.9 billion, and then fell every year until 1982 to $24.4 billion. The estimates prepared by the President's Council on Environmental Quality (CEQ) are shown in Table 1. CEQ estimates the nation spent $36.9 billion in 1979 to comply with Federal environmental protection regulations. (This spending, necessitated by Federal regulations, is referred to as 'incremental' expenditure.) The $36.9 billion amounted to approximately 1.5% of GNP in 1979 [16 (p. 393)]. CEQ estimates that this spending in the future years is likely to grow. Table 1 shows that by 1988, expenditures necessitated by Federal environmental measures will grow to $69.0 billion (in 1979 dollars).

Third, 'indirect' costs can arise because environmental regulations induce firms to seek less than 'optimal' locations or to keep and maintain older, less economical, plant and equipment in order to stay in present locations.

4. PRODUCTIVITY AND ENVIRONMENTAL REGULATIONS

Economic performance is gauged by statistical indicators that reflect changes in both output (goods and services) and input (labour and capital). Productivity simultaneously captures both aspects of economic performance. Inclusion of both cost reduction and quality improvements as important parts of productivity is also noted. Productivity, in general, can simply be defined as an output per unit of input. While there is considerable diversity of opinion about the different items included in the output and input, the most common procedure is to measure productivity by obtaining an estimate of final aggregate private-sector output divided by the number of worker hours of labour input [8 (p. 56)]. However, this definition does not reflect the full set of inputs that are normally used in producing goods and services. Due to this weakness, efforts have been made to construct more comprehensive productivity indicators — e.g. private-sector output per total factor input [8 (p. 56)].

Irrespective of the indicator employed, productivity performance in the United States has been far weaker in the 1970s and early 1980s than in the 1960s. A graph presented in *Civil Engineering* [17] showed US and foreign construction productivity from 1970 to 1979. The article stated that of the world's industrial nations, only the

Table 1 — Estimated incremental pollution abatement expenditures,[a] 1979–88 [16 (p. 394)] (billions of 1979 dollars)

Programme	1979			1988			Cumulative (1979–88)		
	Operation and maintenance	Annual capital costs[b]	Total annual costs	Operation and maintenance	Annual capital costs[b]	Total annual costs	Operation and maintenance	Capital costs[b]	Total costs
Air pollution									
Public	1.2	0.3	1.5	2.0	0.5	2.5	15.8	3.7	1.5
Private									
Mobile	3.2	4.9	8.1	3.7	11.0	14.7	32.1	83.7	115.8
Industrial	2.0	2.3	4.3	3.0	4.1	7.1	25.8	33.0	58.8
Electrical utilities	5.5	2.9	8.4	7.6	5.7	13.3	62.3	42.7	105.0
Subtotal	11.9	10.4	22.3	16.3	21.3	37.6	136.0	163.1	299.1
Water pollution									
Public	1.7	4.3	6.0	3.3	10.0	13.3	25.1	59.2	84.3
Private									
Industrial	3.4	2.6	6.0	5.4	4.5	9.9	42.0	34.0	76.0
Electric utilities	0.3	0.4	0.7	0.3	0.9	1.2	2.9	6.5	9.4
Subtotal	5.4	7.3	12.7	9.0	15.4	24.4	70.0	99.7	169.7
Solid waste									
Public	<0.05	<0.05	<0.5	0.4	0.3	0.7	2.6	2.0	4.6
Private	<0.05	<0.05	<0.5	0.9	0.7	1.6	6.4	4.4	10.8
Subtotal	<0.05	<0.05	<0.5	1.3	1.0	2.3	9.0	6.4	15.4
Toxic substances	0.1	0.2	0.3	0.5	0.6	1.1	3.6	4.6	8.2
Drinking water	<0.05	<0.05	<0.05	0.1	0.3	0.4	1.3	1.4	2.7
Noise	<0.05	0.1	0.1	0.6	1.0	1.6	2.6	4.3	6.9
Pesticides	0.1	<0.05	0.1	0.1	<0.5	0.1	1.2	<0.5	1.2
Land reclamation	0.3	1.1	1.4	0.3	1.2	1.5	3.8	11.5	15.3
Total	17.8	19.1	36.9	28.2	40.8	69.0	227.5	291.0	518.5

[a] Incremental costs are those made in response to Federal legislation beyond those that would have been made in the absence of that legislation.
[b] Interest and depreciation.

US has seen construction productivity drop in recent years. The article further cautioned that some of this may be due to the exclusion in the US data of prefabricated elements in the constructed projects.

We need to explore items that contribute to this drop in productivity. Clearly those items that increase the marginal input required or decrease the output would adversely affect productivity. In a general sense, Haveman has suggested that items contributing to drop in productivity growth relate to composition of output, composition of the labour force, capital–labour ratio, energy prices, lack of sufficient investment in research and development and government regulations, including environmental regulations [8 (pp. 57–59)].

The focus of this article is on the role environmental regulations play in this drop in productivity. Denison, as cited by Haveman [8 (p. 64)], studied the role of environmental regulations on productivity. In his study, it was assumed that the inputs required for environmental control are diverted directly from marketable outputs, resulting in an equivalent decrease in the output numerator of the productivity index. In addition, capital expenditures for pollution abatement were used to reduce measured output by an amount equal to the value of the services this capital would have provided it if were used to produce final products instead of improved environmental quality. The value of these alternative outputs is treated in this analysis as an opportunity cost. Thus, this included both the direct and indirect consequences of expenditure on pollution abatement. Denison estimated the average annual reduction in productivity growth due to environmental regulations for 1973 to 1975 to be 0.22 percentage point, and from 1975 to 1978 to be 0.08 percentage point [8 (p. 66)]. In his comprehensive study on the effect of environmental regulations on productivity he concluded '... regulations still appear to account for a relatively small portion of the measured productivity slowdown' [8 (p. 67)].

A comprehensive macroeconomic study about the possible effects of investment in pollution control on productivity, for the period 1978 to 1986, was conducted by Data Resources Incorporated (DRI). This analysis suggests that over the entire period, productivity growth is estimated at an average of 0.1 percentage point less due to investment in pollution control [8 (p. 70)].

It is clear that environmental regulations do affect productivity in terms of goods and services produced. A drop in productivity of 0.1% or more is still a significant issue. A general climate of uncertainty created by environmental regulations and delays caused by the administration of regulations may have significant indirect effects on other elements contributing to a drop in productivity (for example, capital-labour ratio, energy prices, and an unfavourable international competitive position).

5. APPROACHES TO ENVIRONMENTAL REGULATION AND ADMINISTRATION

Approaches to environmental regulations such as command-and-control economic incentives, and regulatory budget are discussed here.

5.1 Command-and-control
Most air and water pollution control regulations in the United States have relied on technology-based emission standards. In the technology-based approach, the abate-

ment regulations for an industrial category are determined primarily by considering the abatement technologies available and are applied uniformly to all plants in the category [18 (p. 106)]. Given their inflexible requirements for industrial compliance these types of emission controls are often called 'command-and-control' environmental regulations. 'These standards were supposed to reflect the best performance in commercial operation at the time of promulgation, and were to be revised every few years' [18 (p. 106)].

5.2 Economic incentives

Several proposals to regulate environmental pollution involve instituting various economic incentive schemes.

- Probably, the most efficient method of environmental regulation from an economic point of view is through a system of effluent fees. Under this programme, each polluting firm must pay a fee or tax for each unit of pollutant discharged [18 (pp. 127–128)] As a result, both the producer (through the tax) and the consumer (through subsequent price increases) are forced to internalize the pollution externality. This has the advantage of forcing the manufacturer to either change his level of production or adopt some pollution abatement technology. It also forces the consumer to decide whether to consume less of the good or service. But there are a number of technical problems in continuously monitoring different pollutant concentrations and in determining the appropriate fee structure for the pollutants.
- Recently, several proposed approaches to environmental regulations involving negotiable effluent discharges permits have been advanced. Actually, these are not very different from the effluent charges, except that the fees are set by some type of market mechanism. Under these schemes, area-wide concentration limits are set for various types of pollutants. Then permits that specify certain pollution limits are assigned to the existing firms. In addition, agencies may actually set up a bargaining board to determine the negotiated values for these permits. These permits are then considered to be property that can be traded or saved.
- Several writers and analysts have proposed that environmental regulations can be handled by negotiation between the injured parties and those responsible for the polluting activity; that is, if the injured parties are assigned property rights which are legally enforceable. This is a direct analogy of Ronald Coase's idea that the injured parties of an externality may be able to negotiate with either party of the exchange where the externality is generated [6]. An interesting application of this principle in Great Britain is described by Dales [19] in which the Anglers' Cooperative Association was legally given property rights to a trout fishing stream on Royal property. Although various industrial firms (including an electric utility that produced warm effluent) and cities used the stream for their wastes, the Anglers' Cooperative Association was able to maintain water quality (through legal envorcement) that was good enough for the trout to prosper.

5.3 Regulatory budget

In a 1979 discussion paper, Demuth presented an innovative alternative to regulatory reform in the form of a proposal to establish a 'regulatory budget'. Under this approach, the President of the United States and Congress, in a manner similar to that of the annual fiscal budget, would establish prior limitations on the total cost which Federal regulatory agencies could impose on the economy. This approach, it is suggested, would permit elected political officials to decide explicitly what portion of the economy's product is to be devoted to regulatory activity. This discussion paper stated that while the approach has considerable theoretical appeal, it suffers from a number of practical difficulties. For example, just the task of collecting and analysing the necessary cost information would be a major undertaking.

6. INEFFICIENCIES AND SUGGESTED REMEDIES

In order to focus on practices for the administration of regulations that achieve complicance with minimal direct cost, lost time, and other undesired side-effects, it is postulated that problems associated with environmental regulations arise largely from two types of inefficiencies: *allocative inefficiencies* and *x-inefficiencies*. These inefficiencies are examined, and challenging and specific recommendations for eliminating them are presented.

6.1 Allocative inefficiencies

Allocative inefficiency arises when resources are used to make the wrong product [21 (p. 816)]. In the case of environmental regulations, this arises primarily in the regulation formulation process. Participation by engineers and scientists knowledgeable in the area at different policy development levels would be most helpful. Most of us understand that public policy affects the extent to which resources are available for clean water, clean air, environmental protection, and major construction projects, as well as resources for developing technology necessary for achieving these goals. We often ignore the effect technology can have on policy. It is essential that knowledgeable engineers and scientists be involved in the trade-offs between conflicting demands on resources and public policy decisions.

Recommendations for minimizing allocative inefficiencies could focus on means to:

- Eliminate duplicative state and federal regulations.
- Minimize burdensome administrative requirements for compliance with environmental regulations.
- Establish a mechanism so that the costs associated with environmental regulations are weighed against the benefits they might generate.
- Perform an economic analysis where marginal costs for more stringent environmental regulations are compared with the additional benefits they might generate.
- Eliminate unnecessary paperwork, delays, and uncertainty associated with compliance with environmental regulations.

6.2 X-inefficiencies

X-inefficiency arises when resources are used in such a way that even if the right product is being made less productively than is possible [21 (p. 816)]. This means that those persons responsible for incorporating environmental requirements in the project implementation process (engineers, planners, and other decision-makers) need to find more productive and innovative techniques to minimize costs associated with regulatory compliance.

Considerable room exists for improvements in this case. As stated by the GAO report [22], many Federal agencies 'comply with environmental regulations in an uneconomical and ineffective manner because they do not perform evaluations to determine the most appropriate way to achieve regulatory compliance'.

To focus on elimination of x-inefficiencies, some suggestions are:

- To achieve compliance more economically and effectively, engineers, planners, and other decision-makers need to evaluate compliance alternatives and start the compliance procedures early in the planning stages.
- In a similar manner, decision-makers need to improve management and supervision practices to accommodate regulatory requirements. Essentially, responding to new regulatory requirements is not altogether different from accommodating new technology inputs.
- Engineers must develop new skills, where necessary, to implement environment-related requirements and must also develop the necessary management procedures to ensure that environmental compliance is obtained in an efficient and effective manner.
- New technology and an investment in research and development can also play a major role in eliminating these inefficiencies.

7. CONCLUSIONS

This chapter examines several public policy issues concerning environmental regulations, effects of environmental regulations on the economy; and various approaches to environmental regulation and administration. In order to focus on practices for administration of environmental regulations that achieve compliance with minimal direct cost, lost time, and other undesired side effects, it is proposed that problems associated with environmental regulations arise largely from two types of inefficiencies: allocative inefficiencies and x-inefficiencies. These inefficiencies are discussed and specific recommendations for minimizing them are presented.

REFERENCES

[1] Kneese, A. V. *Economics and the Environment*, New York, Penguin, 1977.
[2] Von Neumann, J. Uber ein Okonomisches Gleichungssystem und eine Verallgeneirerung des Brouwerschen Fixpunktstzes, *Ergebnisse eines Mathmatischen Kolloquiums*, **8** (1937), pp. 73–83; English translation, 'A Model of General Equilibrium', *Review of Economic Studies*, **13**, (1945–1946), 1–9.
[3] Arrow, K. J. and Debreu, G. Existence of an equilibrium for a competitive economy. *Econometrica*, **22** (1954), pp. 265–290.
[4] Debreu, G. *Theory of Value: An Axomatic Analysis of Economic Equilibrium*. London: Yale University Press, 1959.

[5] Quirk, J., Saposnikm R. *Introduction to General Equilibrium Theory and Welfare Economics*. New York: McGraw-Hill, 1968.

[6] Coase, R. The problem of social costs, in *Economics of the Environment*, 2nd edn, pp. 142–171. R. Dorfman and N. S. Dorfman (eds). New York: W. W. Norton & Company, New York, 1977.

[7] Solow, R. M. The economics of resources or the resources of economics, in *Economics of the Environment*, 2nd edn, pp. 354–370. R. Dorfman and N. S. Dorfman (eds). New York: W. W. Norton & Company, 1977.

[8] Haveman, R. H. and Christainsen, G. B. Environmental regulations and productivity growth, in *Environmental Regulation and the U.S. Economy*, pp. 55–76. H. M. Peshin, P. R. Portney, and A. V. Kneese (eds). London: Johns Hopkins University Press, London, 1981.

[9] Council on Environmental Quality, Executive Office of the President. *News Release* (9 October, 1980).

[10] Shabecoff, P. Poll shows environmental protection law support. *New York Times* reported in *Champaign-Urbana News Gazette*, 8 October 1981.

[11] Water Pollution Control Federation. *Highlights*, 20 (January, 1983).

[12] Jain, R. K., Urban, L. V. and Stacey, G. S. *Environmental Impact Analysis: A New Dimension in Decision-Making*, 2nd Edn. New York: Van Nostrand Reinhold, 1980.

[13] Ashford, N. and Hill, C. The Benefits of Environment, Health and Safety Regulation. Prepared for the Committee on Government Affairs, U.S. Senate, 96th Congress, 2nd Session (25 March, 1980).

[14] Litan, R. E. and Nordhaus, W. D. *Reforming Federal Regulation*, London: Yale University Press, London, 1983.

[15] Farber, K. D., Dreiling, F. J. and Rutledge, G. L. Pollution Abatement and Control Expenditures, 1972–82, *Survey of Current Business*, **64** (February 1984), 22–30.

[16] 'Environmental Quality'. Eleventh Annual Report of the Council on Environmental Quality, US Government Printing Office, Washington, D.C., 1980.

[17] 'Productivity in Civil Engineering and Construction'. *Civil Engineering*, ASCE, Feb., 1983.

[18] Harrington, W. and Krupnick, A. J. Stationary source pollution and choices for reform, in *Environmental Regulation and the U.S. Economy*, pp. 105–130. H. M. Peshin, P. R. Portney and A. V. Kneese (eds). London: The Johns Hopkins University Press, London, 1981.

[19] Dales, J. H. The property interface, in *Economics of the Environment* 2nd edn, pp. 172–186. R. Dorfman and N. S. Dorfman (Eds.), W. W. Norton, New York, 1977.

[21] Lipsey, R. G. and Steineer, P. O. *Economics*, 5th edn, Harper and Row, New York, 1978, p. 816.

[22] GAO Report to the Congress, *Federal Water Resources Agencies Should Assess Less Costly Ways to Comply with Regulations*, CED-81-36, 17 Feb., 1981, pp. 11, 42.

BIBLIOGRAPHY

Anderson, F. R., Kneese, A. V., Reed, P. D., Serge, and Stevenson, R. B. *Environmental Improvement Through Economic Incentives.* London: Johns Hopkins University Press, London, 1977.

Freeman, A. M. III. *The Benefits of Environmental Improvements: Theory and Practice.* London: Johns Hopkins University Press, 1979.

Mishan, E. J. *Cost–Benefit Analysis,* London: Praeger Publishers, 1976.

Nadiri, M. Ishaq. Some approaches to the theory and measurement of total factor productivity: a survey, *Journal of Economic Literature*, **8**, (December 1970), 1137–1177.

Schelling, T. C. The life you save may be your own, in *Problem in Public Expenditure Analysis*, pp. 127–162. S. B. Chase, Jr (ed.). Washington, D.C.: The Brookings Institution, 1968.

Appendix
List of participants

Dr L. W. Canter, Environmental and Ground Water Institute, University of Oklahoma, 200 Felgar Street, Norman, Oklahoma 73019, USA.
Theme: Hazardous/Toxic Waste
Paper title: Technologies for Treatment and Management of Selected Hazardous/Toxic Wastes

Dr A. Clark and **Prof. R. Perry**, Public Health Engineering, Imperial College of Science and Technology, South Kensington, London, England.
Theme: Hazardous/Toxic Waste
Paper title: Cement-based Stabilization/solidification Processes for the Disposal of Toxic Waste

Dr B. D. Clark, Centre for Environmental Management and Planning, University of Aberdeen, Department of Geography, High Street, Old Aberdeen AB9 2UF.
Theme: Environmental Policy and Regulations
Paper title: The EEC Directive on Environmental Assessment and its Effect on UK Environmental Policy

Dr G. Comati, US Army Research, Development and Standardization Group — UK, 223–231 Old Marylebone Road, London NW1 5TH, England.

Dr F. J. Colon, Department of Physical and Chemical Technology, TNO, PO Box 342, 7300 AH Apeldoorn, The Netherlands.
Theme: Hazardous/Toxic Waste
Paper title: Hazardous Waste: Status and Future Trends of Treating Technologies

Appendix — List of participants

Dr D. R. Cope, Environmental Team Leader, IEA Coal Research, 14/15 Lower Grosvenor Place, London SW1W 0EX.
Theme: Air Pollution Control
Paper title: Coal Combustion and Emission Regulations: European Experiences and Expectations

Dr J. D. Day, US Army Research, Development and Standardization Group — UK, 223–231 Old Marylebone Road, London NW1 5TH, England.

Professor R. I. Dick, School of Civil Environmental Engineering, Cornell University, Hollister Hall, Ithaca, NY 14853, USA.
Theme: Waste Water Treatment
Paper title: Wastewater Treatment and Management of Resulting Residues

Dr F. Fiessinger and Dr Mallevialle, Société Lyonnaise des Eaux et de l'Eclairage, Laboratoire Central, 38 rue de President Wilson, 78230 le Pecq, France.
Theme: Water Treatment
Paper title: Megatrends in Water Treatment Technologies

Dr H. Hahn, Institut fur Diesling Swasserwirtschaft, Universitat Karlsruhe, West Germany.
Theme: Waste Water Treatment
Paper title: Wastewater Treatment with Coagulating Chemicals

Professor K. J. Ives, Department of Civil and Municipal Engineering, University College London, Gower Street, London WC1E 6BT.

Dr R. K. Jain and D. Robinson, Environmental Division, USA-CERL, PO Box 4005, Chapaign, Illinois 61820, USA.
Theme: Environmental Policy and Regulations
Paper title: Environmental Regulation Formulation and Administration

Dr R. K. Jain and Dr L. W. Canter (address as above)
Theme: Hazardous/Toxic Waste
Paper title: Policies and Regulations Affecting Hazardous Toxic Wastes

R. Macrory, Centre for Environmental Technology, Imperial College of Science and Technology, 48 Prince's Gardens, London SW7 1LU.
Theme: Environmental Policy and Regulations
Paper title: Controlling Pollution — Contemporary Issues from a Legal Perspective

J. E. Manwaring, American Waterworks Association, Research Foundation, 6666 W Quincy Avenue, Denver, Colorado 80235, USA.
Theme: Water Treatment
Paper title: Emerging Technologies in the Water Supply Industry

Dr. D. H. Marks and Dr D. Thurston, Department of Civil Engineering, Massachusetts Institute of Technology, Cambridge, Massachusetts 02139, USA.
Theme: Environmental Policy and Regulations
Paper title: Policies and Regulations Affecting Water and Wastewater Treatment Practices

Dr H. J. Oels, Federal Environment Agency, Bismarkplatz 1, D-1000 Berlin 33, FRG
Theme: Air Pollution Control
Paper title: Technical and Regulative Measures for Air Pollution Control in the Federal German Republic

Dr R. Primus, College of Business Administration, Wright State University, Dayton, Ohio 45435, USA

Professor S. P. Shelton, Department of Civil Engineering, Univeristy of New Mexico, Albequerque, NM, 87131, USA.

Dr J. Siboney, Anjou Recherche, Centre de Recherche de la Compagnie Générale des Eaux, Chemic de la Digue, BP 76 — 78600 Maisons Lafitte, France.
Theme: Wastewater Treatment
Paper title: Waste Water Treatment by Fixed Films in Biological Aerated Filters

Dr V. L. Snoeyink, 3230 Nathan M Newmark Civil Engineering Laboratory, University of Illinois at Urbana, 208 North Romine Street, Urbana, Illinois 61801, USA
Theme: Water Treatment
Paper title   Drinking water treatment technology

Dr J. J. Stukel, Vice Chancellor for Research, 1737 W. Polk St., Box 6998, Chicago, IL, 60680, USA.
Theme: Air Pollution Control
Paper title: Abatement Technologies for Air Pollutants

Dr I. Suffet, Environmental Studies Institute, Drexel University, Philadephia, Pennysylvania 19014, USA.

Appendix — List of participants

Dr Summerer, Umweltbundesamt, Bismarkplatz 1, D-1000 Berlin 33, Germany.

Theme: Environmental Policy and Regulations
Paper title: Environmental Regulation and Political Programming

Dr T. H. Y. Tebbutt, University of Birmingham, Department of Civil Engineering, PO Box 363, Birmingham B15 2TT. Telephone: 021 472 1301.

Theme: Water treatment
Paper title: Water Treatment Research and Development — a European Viewpoint

Dr I. M. Torrens, OCDE, Resources and Energy Division, 2 Rue Andre Pascal, 75775 Paris, France.

Theme: Air Pollution Control
Paper title: Economic Issues in the Control of Air Pollution from Fossil Fuel Combustion

Index

abatement technologies for air pollutants, 42–60
    control of gaseous emissions, 51
    control of nitrogen oxide emissions, 55
    control of particulates, 43
    control of VOC, 57
    simultaneous removal of NO and SO, 58
adsorption, 104
air pollution
    economic issues, 61–73
Air stripping, 105

barrier concept, 110
biological aerated filters, 170–175
    operating principle, 170
    results on urban sewerage, 171
    advantages of 173
    operating experience, 173
biological stability, 100

chemical treatment (of wastewater), 145–169
chlorine dioxide, 103

dust emission reduction, 37

economic issues (in the control of air pollution), 61–73
    benefits of air pollution control, 72
    costs of control, 65
    costs of emissions reduction, 68
    criteria for policy action, 73
    economic issues, 69
    emission standards (OECD), 62
    major air pollution issues, 61
    technologies, 63
emerging technologies in water supply, 84–96
environmental assessment, 225–236
    EC directive, 223
    UK project authorization, 226
    UK procedure, 228
environmental policy formulation, 251–256
    counterproductive policies, 253
    proposed solutions, 255
    response of users and generators, 252
    traditional approaches, 252

environmental regulation, 257–269
    approaches to, 264
    benefits of, 261
    concerns regarding, 259
    costs of, 261
    economic incentives, 265
    inefficiencies and remedies, 266
    markets (failure of), 258
    pareto optimality, 258
    productivity and, 262
    public view of, 260
    regulatory budget, 265

Federal Republic of Germany (air pollution control in), 17–41
    federal immission control law, 17
    mobile sources, 38
    reduction in dust emissions, 37
    reduction of SO emissions, 26
    reduction of NO emissions, 33

granular activated carbon, 108–110

hazardous toxic/wastes (technologies for treatment and management), 179–197
hazardous/toxic wastes (status and future trends), 198–222
    definitions of, 179, 198
    management of, 203
    reuse of, 205
    treatment costs, 206
    waste disposal, 193
    waste reduction technologies, 180
    waste treatment and ddisposal, 187

injection wells, 194

megatrends in water treatment technolgies, 77–83
monochloramine, 97

NO emission reduction, 33

ozone, 101

Index

pollution control (UK legal perspective), 237–250

Regulatory measures for pollution control (FRG), 17–41

SO emission reduction, 26

UK pollution control (legal perspectives), 237–250
  best practicable environmental option, 243
  EC directive, 240
  environmental assessment, 240
  enforcement, 246
  information, 245
  integration, 243
  participation, 244
  UK planning system, 238

waste reduction technologies, 180
waste treatment, 187
  chemical stabilization, 187
  thermal treatment, 187
wastewater treatment (research needs), 133–143
  chemical treatment, 145
  destruction of pathogens, 136
  land application of sludge, 141
  microcontaminants, 136
  nutrient removal, 136
  removal of biodegradable organic compounds, 135
  sludge combustion, 140
  sludge reclamation, 140
  stabilization, 139
  suspended solids removal, 135
wastewater treatment (by fixed films in biological aerated filters), 170–175
wastewater treatment with coagulating chemicals, 144–169
  average quality of liquid phase, 145

chemical treatment, 160
comparison of costs, 153
cost of chemical treatment, 149
cost of chemicals, storage and dosing, 150
cost of liquid–solid separation, 151
efficiency and cost, 145
experience of, 156
future development, 164
plant efficiency data, 146
water quality
  aesthetic aspects of, 118
  assessment of, 118
  iron and manganese, 89
  standards and guidelines, 120
water supply (emerging technologies, 84–96
  automation, 87
  corrosion, 90
  disinfection, 87, 97
  monitoring, 88
water treatment, 77–129
  adsorption, 104
  air stripping, 105
  automation, 81
  barrier concept, 110
  biological treatment, 80, 89
  coagulation, 123
  chlorine dioxide, 103
  clarification, 78, 123
  developments in, 123
  disinfection, 97, 125
  filtration, 124
  granular activated carbon, 108
  micropollutant removal, 126
  operation and control, 127
  oxidation, 79, 97
  ozonation, 101
  sludge treatment and disposal, 127
  softening, 81
  volatile organic carbon removal, 105